AF698407

SAFE AND SUSTAINABLE ARCTIC SHIPPING MANAGEMENT AND DEVELOPMENT

SAFE AND SUSTAINABLE ARCTIC SHIPPING MANAGEMENT AND DEVELOPMENT

MAWULI AFENYO

Department of Maritime Business, Texas A&M University, Galveston, TX, United States

ADOLF K.Y. NG

Faculty of Business and Management, BNU-HKBU United International College, Zhuhai, Guangdong Province, China

NAIMA SAEED

School of Business and Law, University of Agder, Kristiansand, Norway

ELSEVIER

Elsevier
Radarweg 29, PO Box 211, 1000 AE Amsterdam, Netherlands
125 London Wall, London EC2Y 5AS, United Kingdom
50 Hampshire Street, 5th Floor, Cambridge, MA 02139, United States

Copyright © 2024 Elsevier Inc. All rights are reserved, including those for text and data mining, AI training, and similar technologies.

Publisher's note: Elsevier takes a neutral position with respect to territorial disputes or jurisdictional claims in its published content, including in maps and institutional affiliations.

No part of this publication may be reproduced or transmitted in any form or by any means, electronic or mechanical, including photocopying, recording, or any information storage and retrieval system, without permission in writing from the publisher. Details on how to seek permission, further information about the Publisher's permissions policies and our arrangements with organizations such as the Copyright Clearance Center and the Copyright Licensing Agency, can be found at our website: www.elsevier.com/permissions.

This book and the individual contributions contained in it are protected under copyright by the Publisher (other than as may be noted herein).

Notices
Knowledge and best practice in this field are constantly changing. As new research and experience broaden our understanding, changes in research methods, professional practices, or medical treatment may become necessary.

Practitioners and researchers must always rely on their own experience and knowledge in evaluating and using any information, methods, compounds, or experiments described herein. In using such information or methods they should be mindful of their own safety and the safety of others, including parties for whom they have a professional responsibility.

To the fullest extent of the law, neither the Publisher nor the authors, contributors, or editors, assume any liability for any injury and/or damage to persons or property as a matter of products liability, negligence or otherwise, or from any use or operation of any methods, products, instructions, or ideas contained in the material herein.

ISBN: 978-0-443-18849-7

For information on all Elsevier publications
visit our website at https://www.elsevier.com/books-and-journals

Publisher: Mica H Haley
Acquisitions Editor: Kathryn Eryilmaz
Editorial Project Manager: Debarati Roy
Production Project Manager: Gomathi Sugumar
Cover Designer: Vicky Pearson Esser

Typeset by STRAIVE, India

ELSEVIER Book Aid International
Working together to grow libraries in developing countries
www.elsevier.com • www.bookaid.org

Dedication

For Dr. David G. Barber
Pearls are everywhere but not the same as the eyes
Thank you.

Contents

Section B Economic opportunities and risks

Section C Operational challenges of Arctic shipping

Photo collection

A photo illustrating a ship trapped in the frozen sea ice during the winter in northern Canada. *(Source: Authors, taken in February 2018.)*

Extreme physical conditions substantially increase the difficulty in the construction, operation, and maintenance of transport infrastructures in the Arctic region. This is a photo illustrating the typical physical landscape in the Arctic region. *(Source: Authors taken in October 2017.)*

Ice melting in the summer has made the Arctic passages more accessible to shipping activities. However, as illustrated in this photo, sea ice does not always melt completely but often ends up with small, numerous icebergs flowing along the Arctic seas and passages, posing considerable risks to the ships that attempt to navigate through. *(Source: Courtesy of Thomas Linkowski from Amundsen Science Program, August 2019.)*

A photo illustrating the CCGS *Amundsen* navigating through the Arctic waters with splintered sea ice in front of it. *(Source: Tommy Pontbriand from Amundsen Science Program, August 2019.)*

The hazards of Arctic shipping are not only restricted to sea ice. This photo illustrates a ship navigating through the Northwest Passage with thick fog and thus poor visibility. *(Source: Adam Tremblett from Amundsen Science Program, August 2019.)*

A photo illustrating the view of the frozen sea near Greenland from a ship's engine control room. *(Source: Source: Petermann Glacier from Amundsen Science Program, August 2019.)*

A photo illustrating the natural landscape along the Northwest Passage in the summer. *(Source: Julia Baak from Amundsen Science Program, August 2019.)*

A photo illustrating the Arctic natural landscape in the summer in Greenland. *(Source: Petermann Glacier from Amundsen Science Program, August 2019.)*

A photo illustrating the Arctic natural landscape in the summer in Greenland. *(Source: Petermann Glacier from Amundsen Science Program, August 2019.)*

A photo illustrating two researchers conducting their work in the CCGS *Amundsen*. *(Source: Petermann Glacier from Amundsen Science Program, August 2019.)*

Inland connection plays a significant role in affecting the success of Arctic shipping. This photo illustrates the damaged rail tracks connecting the town of Churchill with the rest of Canada due to floods. The damaged infrastructures took more than a year to repair, causing huge socio-economic consequences to local communities. *(Source: Authors, taken in October 2017.)*

The town of Churchill is a major gateway connecting the Canadian Arctic with the rest of Canada. This photo illustrates the town of Churchill's train station, with the port of Churchill in the background. *(Source: Authors, taken in October 2017.)*

The CCAPPTIA Conference (www.ccapptia.com) involves key scholars, policymakers, and practitioners in addressing Arctic shipping and development. This photo illustrates the opening ceremony of the first CCAPPTIA conference, held in Winnipeg, MB, Canada. *(Source: Authors, taken in May 2018.)*

A photo illustrating a discussion session on Arctic shipping and development during the third CCAPPTIA Conference. It was held in partnership with the Hong Kong Maritime Museum in Hong Kong, China. *(Source: Authors, taken in August 2023.)*

About the authors

Dr. Mawuli Afenyo is currently a professor at Texas A&M University and holds a PhD in engineering. Prior to becoming an academic, he worked in engineering consulting. Mawuli is well published and has authored and coauthored more than 40 journal articles, conference proceedings, and scientific reports. He is also a sought-after speaker at conferences on the Arctic. Dr. Afenyo has taken part in major multidisciplinary, multinational, large-scale scientific projects in different countries around the world, including Finland, Norway, Canada, United Kingdom, Australia, China, Hong Kong, and Ghana. Mawuli is cofounder of the nonprofit organization called CCAPPTIA—An International Forum for Climate Change Adaptation Planning for Ports, Transportation Infrastructure, and the Arctic (see www.ccapptia.com), made up of more than 50 practitioners and researchers around the world. Dr. Afenyo has also been interviewed by different outlets, both print and radio. Some include the Canadian Broadcasting Corporation (CBC) and SciTech Daily, among others, about oil spills. He has also offered his services as a consultant to different government agencies as well as other corporate organizations around the world. Until recently, Dr. Afenyo was the coeditor of the *Maritime Economist*, the official magazine of the International Association of Maritime Economists (IAME).

Adolf K.Y. Ng is currently a chair professor and the acting dean of the Faculty of Business and Management, BNU-HKBU United International College, Zhuhai, China. He received his DPhil from the University of Oxford, excelling in the research and teaching of transport, global supply chains, and green logistics. His scholarly outputs include 9 books, more than 110 papers in leading journals, and other forms of publication. He receives numerous prestigious accolades and frequently offers strategic advice to major organizations, such as the United Nations (UN), the European Commission (EC), and the African Development Bank (AFDB). He is an associate editor of *Maritime Policy & Management*, the chief editor of *The Maritime Economist*, the official magazine of the International Association of Maritime Economists (IAME), and a council member of IAME. He is a cofounder of CCAPPTIA (www.ccapptia.com), an international forum that brings together leading experts around the world in addressing the challenges and opportunities posed by climate change and the Arctic.

Naima Saeed is a professor in supply chain management at the School of Business and Law at the University of Agder, Norway. She has a PhD in logistics from Molde University College (Norway), specializing in maritime logistics. Before joining the University of Agder, Saeed worked as an associate professor in logistics at Molde University College. Saeed was an honorary visiting fellow at La Trobe Business School, Melbourne, Australia, and a visiting scholar at the Robert H. Smith School of Business at the University of Maryland, USA; School of Business IT and Logistics, RMIT, Melbourne, Australia; University of California, Berkeley, United States; Tokyo University of Science, Tokyo, Japan; Nelson Mandela University Business School, South Africa; and at the University of Dar es Salaam Business School, Tanzania. Saeed has published papers in well-acknowledged international peer-reviewed scientific journals and conference proceedings. Saeed received research funding from several sources, including the European Union and the Research Council of Norway. Saeed was a member of the Norwegian Association of Higher Education Institutions' committee for the research methods course (2016–2020).

Preface

Doing research in the Arctic region is never easy or straightforward. Its remoteness and extreme weather conditions seriously restrict the time, seasons, and areas that allow researchers to conduct their work, not to mention the substantial costs often associated before and during the research process (for example, in a three-week research trip conducted by the authors on the research icebreaker CCGS *Amundsen*, the bill read around USD 20,000 per person). In addition, the Arctic region, in many cases, is the home of many Indigenous peoples and communities. According to the World Wildlife Fund (WWF) (www.arcticwwf.org/our-priorities/arctic-communities/), there are more than 40 Indigenous peoples in the Arctic, representing about 10% of the Arctic's total population of about four million. Their very existence has given much food for thought to us when developing the approaches, methods, instruments, and attitudes in Arctic research, given the (increasing) importance to protect the tradition and livelihood of these peoples and communities during the research process. Indeed, in the past decade, we have witnessed many universities and research institutes introducing much more restrictive ethical guidelines on research projects and places where Indigenous people and communities might be affected. Hence, one could easily imagine the challenges and complexity in conducting quality research on virtually any areas and disciplines in the Arctic nowadays.

This is perhaps what has made the contents of this book especially valuable. All of us have our roots from places that (at least seem to) have little connection with the Arctic. Yet, at certain career stages, we have somehow got involved in the study of the Arctic, and we never look back, albeit as an "outsider" it came at a big opportunity cost. In hindsight, we could not imagine how much sweat and sleepless nights that we have committed in conducting fieldwork, writing numerous reports, papers (and, of course, this book), and grant proposals simply to secure enough funds to keep our project ongoing. In addition, we actively organized different workshops and conferences to enhance the impacts of our research, which played a significant role in the establishment of CCAPPTIA, an international forum that brings together leading experts, stakeholders, and right holders from academia, government, industry, consultants, interest groups, and community groups in addressing the challenges and opportunities posed by climate change and the Arctic. Between 2018 and 2023, CCAPPTIA has organized six Arctic-themed workshops and conferences held in Canada, China (including Hong Kong), France, and online. Some of the photos in this book serve as a little reminder of such sweat and sleepless nights. While not necessarily the very first piece of work (although certainly pioneers) that addresses the safety, sustainability, and socio-economic impacts of Arctic shipping and development and likely a long way to achieving perfection, the contents reflect our every bit of effort, commitment, and resilience in contributing to the Arctic and its communities. Published right after having firmly put the COVID-19 pandemic behind us, it will be our greatest satisfaction if readers find the contents interesting and insightful and that the book could contribute to making the Arctic and our world a better place. And what a way to celebrate our decade-old Arctic research by publishing a book on the Arctic with a highly reputable publisher (thanks, Elsevier)!

Needless to say, our research would not be possible without generous support. In this regard, we acknowledge the National Natural Science Foundation of China (#UICR0600041), Genome Canada and Research Manitoba through the GENICE Project (genice.ca), CCAPPTIA (www.ccapptia.com), and Amundsen Science. Some of the photos included in this book were collected during the 2019 Expedition aboard the Canadian research icebreaker CCGS *Amundsen* and made available by the Amundsen Science Program, supported by the Canada Foundation for Innovation through Université Laval. Also, we would like to thank Dr. Gary Stern, the coproject leader of the GENICE Project. His advice, encouragement, and support to us throughout the years have greatly facilitated our research in the Arctic. On a personal note, Adolf K.Y. Ng would like to thank Dr. David G. Barber, who, unfortunately, passed away in April 2022. Being one of Canada's most influential and accomplished Arctic researchers, in 2015, David somehow decided to invite Adolf (who, by then, had minimal Arctic knowledge) to his office, telling him why he was the most appropriate individual to conduct research on the socio-economic impacts of Arctic shipping and convincing him to take up the challenge. The rest, they say, is history.

Finally, in the brave new world of AI technology, we guarantee that not one single word in this book comes from ChatGPT or any other Generative AI.☺

Thank you very much for the support and happy reading!

Mawuli Afenyo
Texas A&M University, Galveston, TX, United States

Adolf K.Y. Ng
BNU-HKBU United International College, Zhuhai, China

Naima Saeed
University of Agder, Kristiansand, Norway

SECTION A

Theoretical settings

CHAPTER ONE

Setting the scene

1 Introduction

This book gives an insight into what is currently happening in the Arctic, especially in shipping and natural resource exploration. By adopting a holistic approach, it presents how Arctic stakeholders and right holders attempt to prepare for the potential windfall to occur in the Arctic area. In addition, the challenges to harnessing the potential of the Arctic are presented. These are geared toward achieving a sustainable and productive Arctic. Finally, it enables both researchers and practitioners to apply the theoretical knowledge obtained in the field to solve challenging Arctic issues. One of its important characteristics is its coverage over different aspects of the Arctic area. These include the legal, financial, economic, and environmental aspects. Also, it closely integrates the implications of Arctic activities with the indigenous way of life. For instance, the indigenous people along the Northwest Passage (NWP), as well as the Northern Sea Route (NSR), have very peculiar characteristics that include the fact that their cultures are highly linked to the Arctic flora and fauna.

We believe that the contents of this book are unique. Publications that are the most closely comparable to this book (e.g., Nordquist et al., 2016; Pongrácz et al., 2020; UNESCO, 2009) largely focus on the legal implications of the Arctic-related activities, environmental resilience, and educational challenges rather than taking a holistic view of the challenges of Arctic-related activities (e.g., responses to potential accidents, development, subsequent management of the Arctic and risks associated with Arctic Shipping and accompanying operations). Indeed, to the best of our knowledge, none of them manage to satisfactorily present the impacts of climate change on the Arctic in such a comprehensive manner. It presents the Arctic in a holistic way that would make it easy for both beginners and those who already have some knowledge on this pristine region. Thus, there is a major research gap on how the Arctic is imparted by shipping, oil, and gas activities negatively, as well as positively. For instance, the huge investments of the

Safe and Sustainable Arctic Shipping Management and Development
https://doi.org/10.1016/B978-0-443-18849-7.00004-X
Copyright © 2024 Elsevier Inc.
All rights are reserved, including those for text and data mining, AI training, and similar technologies.

Russian, French, and Chinese companies in Liquefied Natural Gas (LNG) projects along the NSR have been critical to reviving the fortunes of the Arctic. Furthermore, countries like Canada have recently developed a comprehensive policy that strives to empower the indigenous people within the Arctic area. In this regard, some non-Arctic countries have lobbied for an observer status and tried to influence Arctic policies. For example, the so-called *Silk Road Economic Belt and the 21st Century Maritime Silk Road* (commonly known as the *Belt & Road Initiatives*, or BRI) has, as a critical component, the polar silk road.

The collaborations among highly influential scholars on matters related to the Arctic make this book a rich resource and reference for Arctic-related issues. The knowledge base shows forward-thinking policies of the Arctic by the scholars, predicting a very favorable future of the Arctic. Hence, it will be of great interest to researchers, policymakers, and industrial professionals involved in transport and environmental planning and management (e.g., port authorities, planners, oil spill responders, ship operators, and environmental branch of governmental organizations like the Environmental Protection Agency (EPA) of the United States and Environment Canada. The readers would understand the trend of Arctic development and build the relevant knowledge base and skills to address challenging Arctic problems. For scholars, it is a valuable source for generating new ideas for Arctic research and response to Arctic-related accidents. In this case, in terms of audience, it includes scholars, researchers, and graduate students (e.g., PhD, MPhil, MSc, etc.), as well as senior policymakers and practitioners (e.g., planners, maritime administrators, shipping companies' crewing and/or preparedness managers, marine insurers, maritime consultants, and engineers). For scholars, researchers, and graduate students, it inspires them to generate original, innovative ideas for further research in sustainable transport and regional systems, especially in the Arctic area. For senior policymakers and practitioners, it offers them a comprehensive understanding on the contemporary development and research on Arctic shipping development, planning, and management. Indeed, it presents a unique perspective of Arctic development and can greatly help the facilitation of knowledge transfer and bridges the gap between academic knowledge and professional practice.

Finally, it is noted that this book is a major output of the *Second International Workshop on Climate Change Adaptation Planning for Ports, Transportation Infrastructures, and the Arctic* (CCAPPTIA) (www.ccapptia.com), held simultaneously in Winnipeg (Canada), Marseille (France), and Hong Kong

(China) on 28–30 June 2021. Bringing together more than 100 leading experts and stakeholders from the academia, government, industry, and interest groups, CCAPPTIA aims to consolidate and coordinate global research and development activities that (1) understand the current decision-making process in Arctic shipping, (2) identify attributes that catalyze collaboration between ports and other transport infrastructures, including those in the Arctic, (3) identify ways that can facilitate the deployment and implementation of oil spill response strategies in the Arctic under different geographical and cultural contexts, and (4) assess strategy and policy implementation under the context of Arctic development and management.

2 Chapter layouts

This book consists of 12 chapters categorized under four major sections, namely, Theoretical Settings (this chapter, Chapters 2 and 3), Economic Opportunities and Risks (Chapters 4 and 5), Operational Challenges (Chapters 6–8), Environmental and Social Implications (Chapters 9–11), and Chapter 12 focusing on an emerging topic in terms of cybersecurity. A brief introduction of the proposed chapters can be found in the following.

This chapter, entitled *Setting the Scene*, offers a detailed, holistic overview of the approach to Arctic shipping management and development. As described earlier, it explains the objectives, uniqueness, and contributions of this book. Finally, as illustrated in this section, the chapters of this book are briefly introduced.

Chapter 2, entitled *Arctic Shipping Development and Management: A Holistic View*, discusses the fundamentals of Arctic shipping management and development. This chapter discusses the fundamentals of Arctic shipping management and development. It covers the technical definition of the Arctic, the countries that make up the Arctic, as well as their unique views on the Arctic and shipping in this area. Furthermore, the Arctic Council is also discussed. It raises the fundamental issues currently facing the Arctic, for instance, whether to continue the development of the Arctic or abandon it. Also, it critically discusses the "business view" (as opposed to the "environmental view") of the Arctic area. Furthermore, the chapter presents the latest situation and especially evolutions that took place after 2018 with respect to Arctic development.

Chapter 3, entitled *Arctic Policies, Governance, and Cooperation: A Catalyst for the Development and Management of the Arctic*, examines the national policies of the key Arctic countries, such as Canada, Norway, and Russia.

In addition, the regulations and policies put in place by, for instance, the Arctic Council and the Barents Euro-Arctic Transport Plan are reviewed. In this chapter, we discuss the Polar Code in the context of operations and environmental protection activities in the Arctic area. Furthermore, issues related to Arctic Governance, sustainable development and corporation, infrastructure development, capacity development as well as research related activities are addressed. Finally, it also presents the current structure of the management of Arctic resources by the various Arctic countries.

Chapter 4, entitled *Potential Consequences of Arctic Shipping and Resource Exploitation*, discusses the effect of climate change on Arctic navigation and natural resources exploration. It presents and discusses the implications of resources extraction on the culture, flora, and fauna in this area, as well as the economic benefits in terms of fishing and petroleum. In this case, Rankin Inlet (Nunavut Canada), Greenland (Denmark), and Svalbard (Norway) are used as the case studies to address the implications of navigation and resource exploration in the Arctic area. In addition, consequences of resource exploration in countries like United States, Finland, Russia, and Iceland are also discussed. Finally, the impact of carbon emission is also analyzed.

Chapter 5, entitled *Operational Risks of Arctic Shipping*, presents and discusses the risk factors during operations in the Arctic. The different types of risk assessment methods available to assess operational risks are discussed, and one of them is demonstrated. Based on the work presented, we found that, in general, operational risks currently outweigh the readiness of the Arctic communities and ship operators to address those risks. Furthermore, it addresses the role of the Polar Code for safe Arctic shipping. Finally, the perspective of the professionals on its implementation, as well as its success and challenges are discussed.

Chapter 6, entitled *Arctic Oil Spills: Should We Be Worried?* discusses the consequences of potential oil spills by ships navigating the Arctic area. The question about if we should worry about oil spills in the Arctic is addressed in a comprehensive way. In addition, the need to begin to examine similar consequences for oil and gas development is presented. Many types of accidents can potentially be generated by Arctic shipping, not least oil spills. A comprehensive review of implications of past oil spill incidents and consequences is analyzed. Furthermore, we critically discuss the policy implications of potential oil spill problems for the Arctic and the need for specific design guidelines for oil and gas equipment which would be developed for the Arctic and deployed and installed in the Arctic. Finally, the use of modern technology, such as drones and Artificial Intelligence (AI), is discussed in the context of oil spill identification and recovery.

Chapter 7, entitled *Arctic Shipping Insurance*, discusses the need for Arctic shipping insurance. A model for predicting ship-specific (including autonomous ships) actuarially fair insurance premium is proposed. The model is a function of the type of ship going to the Arctic. The model is based on the Bayesian theory, coupled with a loss function. The model offers a very useful framework for underwriting insurance for ships navigating within the Arctic area. (Non-)governmental organizations could use the tool for key decision-making purposes. Also, the model has the capability for contingency planning for oil spill response by both the local communities and governmental agencies, such as the coast guards. The experience of a hypothetical insurance company is included as a case study. It is used to analyze the model proposed for Arctic shipping insurance. This has been included in this chapter as an illustrative example of application of the model on the case study.

Chapter 8, entitled *Evaluation of Arctic Resources Exploration, Management, and Supply Chains*, evaluates Arctic resource exploration, management, and supply chains. First, we investigate a link between the resource exploration trends and shipping needs, as is how resource exploration affects shipping development, especially for the Russian Arctic. After then, we discuss the third-party services that support those operations, as well as the resupply of tourism during COVID-19. Finally, we study the risk of supply chain disruption. Like every system, Arctic supply chains are not immune to threats, not least the supply chains that create value for global and local economies.

Chapter 9, entitled *Environmental and Socio-Economic Implications of Arctic Development*, investigates the environmental and socioeconomic implications of Arctic development. The first part examines the environmental impacts of Arctic development. For instance, the possibility of more greenhouse emission because of shipping, oil and gas, and other economic activities. There is a strong case made for the level of reduction of emissions if Arctic shipping is encouraged and is demonstrated by two methods: energy based and activity-based method. In the second part, an influence diagram (ID) is presented to assess the environmental and social impacts of Arctic shipping development. The model is based on the Bayesian theory that estimates the social impacts on human activities on the Arctic, such as a potential oil spill caused by Arctic shipping. In addition, Arctic-related social indicators are also discussed to help readers understand how we can measure how well a community is doing in the Arctic.

Chapter 10, entitled *Indigenous Knowledge and Culture: How Important Is it for Arctic Development and Management?* explains the importance of incorporating Indigenous knowledge and culture in Arctic development.

The five factors that are Indigenous knowledge of the Arctic environment, their ability to manage the natural resources sustainably, their cultural heritage, their involvement in policy and decision-making, and recognition of their rights in international laws have been identified as driving forces for integrating their knowledge and culture in Arctic development. The argument is supported with the help of some examples related to Indigenous people in different Arctic nations.

Chapter 11, entitled *The Perspective of Individuals Who Are Involved in Arctic and Climate Change Work*, examines the effects of climate change generally and the relationship with the Arctic. The perspective of professionals working on climate change and Arctic issues are presented. It covers a study that was conducted while the authors were onboard the Canadian Ice Breaker Amundsen as well as the First International Workshop on Climate Change Adaptation Planning for Ports, Transportation Infrastructures, and the Arctic (CCAPPTIA) (www.ccapptia.com). The chapter compiles some of the views and opinions collected from the participants in these exercises. Most of the crew has vast experience in sailing through the Arctic in the past decade and thus possessed immense understandings on the operational risks of Arctic shipping. The views are analyzed to further understand the characteristics and risks that are unique to Arctic shipping which most normal shipping routes do not possess. The CCAPPTIA workshop brought together professionals from almost six continents of the world working on the subject matter.

Chapter 12, entitled *Cybersecurity in the Arctic: Are Businesses and Infrastructures Safe?* examines if the Arctic is ready for a major cyber-attack and if so, how prepared are governments and private companies to deal with such. In addition, the insurance, businesses, and military impacts of cyber is also evaluated for the Arctic region. Finally, the current policies on Arctic maritime cybersecurity are also examined, and proposals are made on how these weaknesses in infrastructure and policy can be addressed. The approach of individual Arctic countries, businesses, communities, and the International Maritime Organization (IMO) is also discussed.

3 Previously published materials

Some of this book's contents have previously been published in academic journals and other forms of publications. They include the following:

- Afenyo, M., Jiang, C., Ng, A. K. Y., & Lee, P. T. (2023). A Bayesian-loss function-based method in assessing loss caused by ship-source oil spills in

the arctic area. *Risk analysis: An Official Publication of the Society for Risk Analysis, 43*(8), 1557–1571. DOI: 10.1111/risa.14025
- Panahi, R., Afenyo, M., & Ng, A. K. Y. (2023). Developing a resilience index for safer and more resilient arctic shipping. *Maritime Policy & Management, 50*(7), 861–875, DOI: 10.1080/03088839.2022.2061059
- Afenyo, M., Ng, A. K. Y., & Jiang, C. (2021). A multi-period model for assessing the socio-economic impact of oil spills in the arctic. *Risk Analysis: An international Journal, 42*(3), 614–633
- Panahi, R., Afenyo, M., Ng, A. K. Y., & Haeri, F. (2020). A novel approach in probabilistic quantification of risks within the context of maritime supply chain: The case of extreme weather events in the arctic. *Accident Analysis & Prevention, 144*, 105673
- Afenyo, M., Ng, A. K. Y., & Jiang, C. (2019). Climate change and Arctic shipping: A method for assessing the impacts of oil spills in the Arctic. *Transportation Research Part D, 77*, 476–490
- Afenyo, M., Khan, F. Veitch, B., & Yang, M. (2017). Arctic shipping accident scenario analysis using Bayesian Network approach. *Ocean Engineering, 133*, 224–230
- Afenyo, M., Khan, F., & Veitch, B. (2016). A state-of-the-art review of fate and transport of oil spills in open and ice-covered water. *Ocean Engineering, 119*, 233–248

References

Nordquist, M., Moore, J., & Long, R. (2016). *Challenges of the changing arctic continental shelf, navigation, and fisheries. Vol. 19*. Center for Oceans Law and Policy.

Pongrácz, E., Pavlov, V., & Hänninen, N. (2020). *Arctic marine sustainability arctic maritime businesses and the resilience of the marine environment* (1st ed.). Springer Polar Sciences.

UNESCO. (2009). *Climate change and arctic sustainable development: Scientific, social, cultural, and educational challenges*. Paris: UNESCO.

CHAPTER TWO

Arctic shipping development and management: A holistic view

1 Definition of the Arctic

The definition of the Arctic can be viewed through different lenses. This may be based on political, geographical, and environmental contexts. The Arctic is mostly described as the most northern part of the earth. In this region, the temperatures are extremely low, and the population of humans is very sparce. Politically, the Arctic, as defined by the Arctic Research and Policy Act of 1984, the specific areas covered for the United States include the whole area that lies north of the Arctic circle, as well as the northwest part of the boundaries formed by Porcupine, Yukon, and Kuskokwim Rivers; the Aleutian chain; and adjacent marine areas in the Arctic Ocean and the Beaufort, Bering, and Chukchi Seas (USDS, n.d.). Many countries challenge the sovereignty of the area of the Arctic and has often resulted in some misunderstanding between countries (Dodds & Nuttall, 2019). For example, the United States and Canada have a serious contention when it comes to the Northwest Passage. The United States asserts that the Arctic is international waters, while Canada maintains it is national waters (Afenyo et al., 2021). This difference often results in divergence in Arctic policy and governance. However, over the years, the two countries have collaborated to deal with Arctic research-related issues (see: https://neba.arcticresponsetechnology.org/about/) as well as natural disasters.

Some scientists define the Arctic geographically as the area in the Arctic circle. The Arctic circle is the latitude 66.5 degrees north of the equator. The majority of the Arctic is ice-covered and includes countries/regions, such as Russia, Canada, Greenland (Denmark), and Alaska (US), while several non-Arctic countries also show substantial interests on the region's development, such as China, India, Italy (High North News, 2023), to name but a few. This is especially true for China, where the Arctic is included as one of the major pillars in its Belt and Road Initiatives—the so-called "Polar Silk Road." See Fig. 1.

Safe and Sustainable Arctic Shipping Management and Development
https://doi.org/10.1016/B978-0-443-18849-7.00010-5
Copyright © 2024 Elsevier Inc.
All rights are reserved, including those for text and data mining, AI training, and similar technologies.

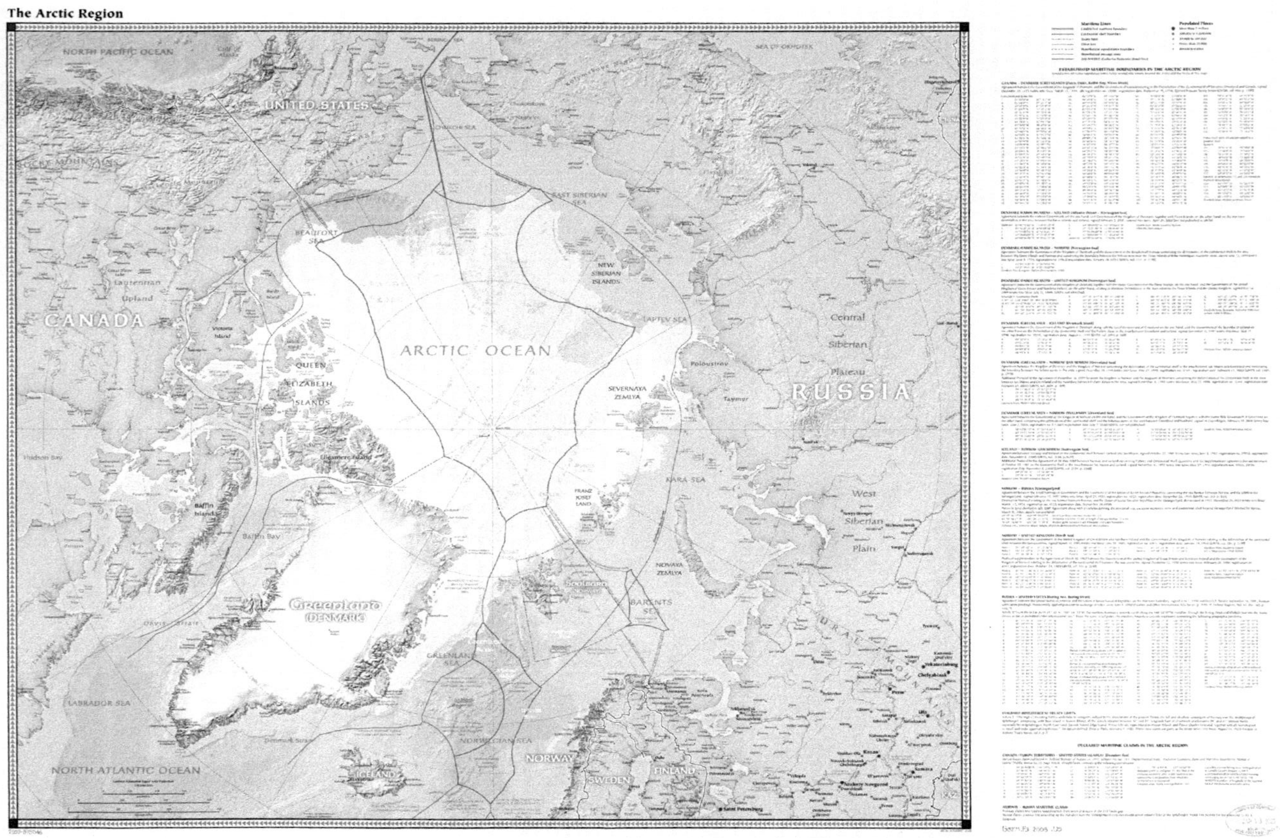

Fig. 1 Definition of the Arctic. *(Courtesy: https://education.nationalgeographic.org/resource/arctic/.)*

2 The Arctic Council

The Arctic Council is the official body that manages and administers policies related to the Arctic. It was established in 1966 and has as its focus promoting intergovernmental cooperation and collaboration among Arctic states (Rottem, 2019). There are eight members in the Council, and any issues and major decisions need to be accepted by all the members before it can be passed. The basis of this group is the Ottawa Declaration. The eight members are Canada, Denmark, Finland, Iceland, Norway, Russia, Sweden, and the United States. In addition, there are six permanent participants that help to shape indigenous policies. These are Aleut International Association, Arctic Athabaskan Council, Gwich'in Council International, Inuit Circumpolar Council, Russian Association of Indigenous Peoples of the North, and the Saami Council. From 1998 to 2006, the following countries obtained observer status: Germany, Netherlands, Poland, UK, France, and Spain. Further observer status was extended to China, India, Italy, Japan, Switzerland, and Singapore between 2013 and 2021. There are also countries with pending observer status, and they need to request their status for each meeting. Some of the countries include Serbia, Turkey, Yemen, Brunei, and Albania.

Also, the key to the Council is the expert and working group. The working groups are six in all, while the expert group is one. One of their major roles is to come up with practices and recommendations that are best for safe as well as sustainable operations in the Arctic region. The working groups are the following: Arctic Contaminants Action Program; Arctic Monitoring and Assessment Program; Conservation of Arctic Flora and Fauna; Emergency Prevention, Preparedness and Response; Protection of the Arctic Marine Environment; Sustainable Development Working Group.

Apart from these groups, there are organizations and countries that hold observer status. However, the chairmanship of the Arctic Council is not permanent, but rotate among the major Arctic countries. The maiden chairmanship was spearheaded by Canada. In addition, each country is represented by a Senior Arctic Official (SAO), who is normally a representative from the foreign affair ministry of the member country. In every 2 years, ministerial declarations are issued, which serve as a guide for the administration and operation of the Arctic Council.

In this regard, it should be noted that even though the Arctic Council recommends policies and guidelines, now it does not have the resources and legal framework to enforce them. In fact, the enforcement lies on the individual member countries. In addition, the Council does not have a budget and relies on individual countries to sponsor projects. Furthermore, they cannot enforce anything militarily; there is no Arctic force.

3 Management and development of the Arctic: The cases of selected countries

The management and development of the Arctic are multifaceted and include various dimensions such as economic, environmental, social, and strategic considerations. In the following, we explain how Norway, Russia, Canada, and the United States are managing and developing the Arctic:

3.1 Norway

For Norway, the Arctic has significant importance from the geopolitical, cultural, and economic point of view. Regarding geographical location, the Northern area constitutes 35% of Norway's land, and by sea, Norway has rights up to 200 nautical miles from its baseline. Norway shares a significant portion (about 80%) of the total traffic, which mainly handles oil and gas extraction and production. Nine percent of the total Norwegian population resides north part of the Arctic Circle. Norway's Arctic policy focuses on safety, strength, and international collaboration (Norwegian Ministry of Foreign Affairs, 2021; Stoltenberg, 2013).

The following are the areas where Norway has contributed to developing and managing the Arctic:

Facilitating International Law: The Law of the Sea regulates all activities in the Arctic Ocean. Norway supports the existing international law. For this purpose, Norway has contributed to several international agreements. For example, Norway is among the five states, including Denmark, Canada, Russia, and the United States, which adopted the Ilulissat Declaration to ensure that the Law of the Sea establishes a guideline to control the activities in the Arctic region. Norway has signed delimitation agreements with Iceland, Denmark (Greenland), and Russia to avoid disputes due to the overlapping of these states' economic zone and continental shelves. In 2019, Norway agreed to a new border agreement with Russia containing revised and updated border details between the two nations by utilizing advanced technology. To protect the Sami people's rights regarding their culture, language, and lifestyle, along with Sweden and Finland, Norway formulated a convent text for Nordic Sami in 2016 (Norwegian Ministry of Foreign Affairs, 2021; Østhagen et al., 2018; Stoltenberg, 2013).

Regional and International Cooperation: Due to its strategic location and growing global interest in the Arctic, Norway has established strong regional and international cooperation (Østhagen et al., 2018). Norway has joined several regional cooperation forums, including the Nordic Council of Ministers, the Arctic Council, the Barents Euro-Arctic Council, the Council of Baltic Sea States, and the Northern Dimension. One of the most critical parts of Norway's Arctic policy is the Barents cooperation. The unique feature of the Barent association, which makes it effective, is the intergovernmental and interregional collaboration as a part of the Barents Euro-Arctic Council and the Barents Regional Council (Norwegian Ministry of Foreign Affairs, 2021).

In 2019, Norway was elected as a chairman of the Barents Euro-Arctic Council for 3 years and contributed significantly to promoting health, integration of people, and knowledge sharing. Canada, Denmark, Finland, Iceland, Norway, Russia, Sweden, and the United States are members of the Arctic Council. In addition to the members' countries, the six organizations represent the indigenous people in the Arctic Council. The Arctic Council makes policies related to climate change, the environment, and sustainability. Norway collaborates with the European Union (EU) in Arctic fisheries, research, and environmental protection. Norway, Sweden, Denmark, and Finland assist the EU in building the Arctic policy. The Norwegian government also decided to contribute to the EU's initiatives for regional territorial collaboration (2021–2027) (Norwegian Ministry of Foreign Affairs, 2021).

Handling Climate Change Challenges: The Arctic is facing new challenges due to climate change. Under the Paris Agreement, the Norwegian government is taking measures to reduce Norway's greenhouse gas emissions (GHGs) by 55% by 2030 and 90%–95% by 2050. Norway is also implementing strategies to decrease its black carbon emission by more than 25% by 2050 (Norwegian Ministry of Foreign Affairs, 2021). Norway has also initiated establishing a Norwegian-Russia collaboration to control marine litter and microplastic in the Arctic. Area-based safety under the Nature Diversity Act is an excellent strategy to protect valuable habitats and ecosystems. In this respect, Norway has shown outstanding progress by safeguarding the 15% of the land area of the two North counties: Nordland and Troms og Finnmark. In North Norway, the government has developed 18 national parks and massive nature reserves, which cover 65% of Svalbard's land to safeguard the Arctic environment (Norwegian Ministry of Foreign Affairs, 2021).

Norwegian government urged Arctic Council countries to emphasize the International Maritime Organization (IMO) to establish a safety guide for vessel activities in the Arctic region. Norway and Russia had convinced the IMO to develop a compulsory vessel reporting procedure for the Barent Sea region. Norway has set high safety and environmental criteria for petroleum activities in the Arctic and is willing to assist other nations in supporting their sustainability efforts (Stoltenberg, 2013).

Managing Marine Resources: The Barents Sea has rich fish resources like the Northeast Arctic cod. Norway and Russia have established the framework of a joint fisheries commission to regulate the fisheries in the Barents Sea. The International Council for the Exploration of the Sea (ICES) assists the commission in developing the policies. Due to the commission's efforts, the quantity of illegally caught fish has significantly declined since 2010. Norway and Russia also collaborated to increase the volume of Codfish, and due to their efforts, its quantity in 2013 was about 10 times higher than it was in 1988. The two countries' joint efforts have resulted in catching fish resources that are more than US$2 billion in value, which is a great food source (Hønneland, 2006; Stoltenberg, 2013).

Economic Development and Growth in North Norway: North Norway has shown tremendous economic growth compared to the rest of the national economy from 2008 to 2018. The government of Norway prioritizes the following industries for further development: ocean-based industries, the maritime sector, petroleum, green power-intensive manufacturing, mineral extraction, agriculture, tourism, space infrastructure, and the services industry. In the Norwegian North, petroleum-related activities have been carried out for decades, and the Barents Sea is full of natural resources (Underhill, 2007). The Norwegian Barents is more suitable for operating conditions than other parts of the Arctic because of the warm North Atlantic current. With its good geographical location and companies with excellent experience and skill in working in harsh environments offshore, Norway has great potential for generating social and economic benefits (Norwegian Ministry of Foreign Affairs, 2021; Stoltenberg, 2013).

The Barents Cooperation contributed to improving the people's living standard in the North by addressing economic growth, environment, education, culture, and integration between people in the North. Previously the region was isolated, especially during the Cold War (Jensen & Rottem, 2010; Stoltenberg, 2013). Due to the combined efforts of the Barents Cooperation member states, the area is now flourishing economically, with good cross-border traffic and a zero-unemployment rate. One of the significant contributions of the Barents Cooperation has been its efforts to improve the relations between the Nordic countries and Russia, including promoting integration among the people of the region (Stoltenberg, 2013).

In North Norway, communities are located far from the economic centers and have a low population density. But despite these obstacles, the business is growing in the region, possessing a high probability of economic growth. But high transport costs and uncertain weather conditions are the main barriers to utilizing business opportunities and convincing people to migrate to North Norway. The government should, therefore, prioritize the improvement and development of infrastructure (Norwegian Ministry of Foreign Affairs, 2021).

A Declining Trend in Population Size: North Norway has experienced a decline in Population since 1970, mainly due to the migration of people to the urban part of the country. In addition to the declining population trend, the percentage of senior citizens is high, while the percentage of children, females, and young labor force is low. This indicates that a higher percentage of the people in North Norway is economically dependent on the young labor force of the region, causing an increasing burden on social security and the public health sector (North & Bullvåg, 2017).

To encourage people to stay in the North and even to migrate from the urban to the North part, the government of Norway has taken several measures, including promoting the plans to provide a good quality of life, job creation, business opportunities, and cultural and leisure programs (Norwegian Ministry of Foreign Affairs, 2021). To address the challenge of the disproportion of elderly and young people in North Norway, North and Bullvåg (2017) suggested the following strategies: to involve older people in economic activities, design national social and health policies to meet the demands of older adults in North which has a higher proportion of the dependent population compared to the national average of Norway, promote companies that provide goods and services to senior citizens, support the art and entertainment sector to engage the older adults in healthy activities, and expand digital health technologies to provide timely and efficient health services to elderly.

3.2 Russia

Russia occupies a vast area of the Arctic that spreads over 24,150 km, covering more than 50% of the Arctic Ocean coastline. The Arctic Circle covers about 20% of Russia's area. The Presidential Decree N296 from May 2, 2014 explained the Russian Arctic as consisting of land and island area in the following eight regions: Murmansk Oblast, Nenets Autonomous Okrug (NAO), Yamalo-Nenets Autonomous Okrug (YNAO), Chukotka Autonomous Okrug, Arkhangelsk region, Komi Republic, the Republic of Karelia, Sakha Republic (Yakutia), and Krasnoyarsk Krai. The first four regions are entirely included in the Russian Arctic, while the others are included partially (Gritsenko & Efimova, 2020).

The total population of the Russian Arctic is two and a half million, which is half of the entire inhabitants of the whole Arctic region. Therefore, developing and managing the Arctic is one of the top areas of interest for the Russian government (See https://arctic-council.org/about/states/russian-federation/).

The following are some critical areas for managing and developing the Russian Arctic:

Legal Framework: In 1921, Soviet Russia's Council of People's Commissars announced a decree that gave the following rights to the Soviet authorities: formulate strategies to protect the environment, fish and animal lands, and operate the Arctic Ocean's commercial areas (Snetkov et al., 2020). The International Treaty of Svalbard ensured that all the Arctic states that joined the Treaty would get the rights to explore the existing and future archipelago's natural resources (Shaw, 2017). The former Soviet Union joined the Treaty in 1935 and got the rights, like other member states, to carry on free economic tasks within the Arctic. The Arctic member states accepted the United Nations Convention on the Law of the Sea (UNCLOS) in 1982. Since its establishment, it has been considered the global legal creed (Snetkov et al., 2020) and determined the rights and duties of the states, including the Arctic states, related to the use and search of marine resources. Russia also follows the framework of international law, primarily the UNCLOS (Kovalev, 2004).

Russia prioritizes national concerns regarding the North Sea, but simultaneously, the Russian Federation is aware of its international responsibilities. For instance, Russia formulated the base for an updated and advanced legal regularization of relationships related to the Northern Sea route by signing the law "On amendments to certain legislative acts of the Russian Federation regarding state regulation of merchant shipping in the water area of the Northern Sea route" (Snetkov et al., 2020). The following services are offered to the vessels in the Russian Arctic regardless of their flag:

- Traffic safety.
- Support to break ice and pilotage.
- Assistance in handling an emergency.
- Access to the coastal infrastructure.

Resource Extraction: The Arctic is rich in natural resources such as oil, gas, minerals, timber, and fish. Russia has access to more than half of the Arctic's natural resources (oil and gas) than the other Arctic states (De Agreda, 2017). Russia has rich mineral resources, including copper, iron, nickel, cobalt, titanium, palladium, platinum, tint, gold, silver, diamond, mercury, antimony, and tungsten. In addition to these minerals, it also has hydrocarbons—coal, oil, and gas (Gritsenko & Efimova, 2020). In 2012, Russia extracted nonferrous and noble metals exceeding 14 million USD, much higher than the United States (3 million USD) and Canada (2 million USD) mined from their region (Bortnikov et al., 2015). However, severe weather and inefficient transport infrastructure make the exploration, progression, and mining of minerals challenging. The mining process requires special techniques and huge investments (Gritsenko & Efimova, 2020). The Russian state-owned company Rosneft is primarily responsible for exploring, developing, and producing Arctic oil and gas. In 2010, Rosneft received four licenses to explore oil and gas in the Arctic. Three out of four licenses are issued for launching the projects in the Kara Sea, while the fourth one is issued for activities in the Pechora Sea. Both sea areas hold about 21.5 billion tons of oil (See https://www.rosneft.com/business/Upstream/exploration/Prospective_projects/arctic_seas/). In July 2022, Rosneft expanded its activities significantly by constructing a new oil terminal at the port of Bukhta Sever. The new terminal development is a part of Rosneft's massive Vostok Oil project. The oil terminal, regarded as Russia's largest terminal, will have a capacity of 102 reservoirs and is expected to tranship 30 million tons of oil annually (https://www.highnorthnews.com/en/russias-rosneft-building-large-scale-arctic-oil-terminal).

Infrastructure Development: Improving the infrastructure and developing a unified Arctic transport system is crucial to utilizing Arctic resources and social and transit traffic developments, which is why infrastructure development is the top priority area in Russian Arctic policy. Russia has been investing in developing infrastructure in the Arctic region. This includes the construction of the world's largest Arctic port of Sabetta on Yamal, the development of an innovative aircraft built especially for polar operation, the renewal of the airpark on Yamal, and the construction of the railway to connect the industrial parts of the Urals and the Arctic areas of Yamal (Serova & Serova, 2019). However, despite these efforts, there is room for improvement to create an effective transport corridor. For instance, there is a need to renovate the Arctic ports in Khatanga, Tiksi, Pevek, Dudinka, and Dixon Island and to build new port resources. To develop the Northern Sea Route up to the standard of the international navigation system, services such as loodsman, icebreaking, maintenance, and repair must be provided at each port (Frolov, 2015).

Environmental Protection: Large industrial projects in the Russian Arctic region to extract oil, gas, and other natural resources negatively impact the environment. The contribution of Russia to GHGs is significant. Russia is well aware of its responsibilities to mitigate its negative influence on climate and has taken several measures, including setting a goal to decrease GHGs to 75% of 1990 size. The Russian government has balanced environmental protection and natural resource exploration, as evident in the country's "Arctic Strategy from 2020." For this purpose, the government's policies highlighted the significance of developing good terms with the neighboring Arctic countries and international cooperation. The Russian government's Arctic strategy highlighted the importance of the Arctic Council and the Barents Euro-Arctic Council in creating a global collaboration among the Arctic countries (https://thebarentsobserver.com/en/2023/04/russia-might-lower-environmental-standards-arctic). Russia was the chairperson of the Arctic Council from 2004 to 2006 and currently holds its second chairpersonship from 2021 to 2023. During its first tenure, Russia emphasized sustainable management of natural resources and promoted international collaboration to protect the environment. As a part of its policy to protect the environment, the Russian government also signed the Kyoto Protocol to the United Nations Framework Convention on Climate Change. For the Kyoto Protocol's implementation, joining the Russian Federation was necessary, and Kyoto Protocol came into force after Russia's accession. In its second tenure, the Russian government emphasizes balancing economic, social, and environmentally sustainable development in the Arctic (https://arctic-council.org/about/states/russian-federation/; https://arcticreview.no/index.php/arctic/article/view/3225/6573#toc).

Indigenous Communities: Russia is home to more than 190 ethnic groups, and people living in the Arctic region are mentioned in the federal law as "Indigenous small-numbered peoples of the North, Siberia, and the Far East of the Russian Federation." The indigenous people of the Arctic are from different communities, such as Sami, Nenets, and Chukchi. They have unique and different cultures, lifestyles, and ways of living. Their source of income depends on traditional economic activities like crafts, reindeer farming, fishing, hunting, gathering, and harvesting marine mammals. The life of the indigenous people in the Arctic region is hard, and they face several challenges due to severe weather conditions and limited access to healthcare and other modern facilities. Russian legislation has been protecting the rights of indigenous people since 1990, and even the law was more constructive for the indigenous people than for the rest of the country (Gorbunova, 2021).

The Russian legal system ensures the following principal rights of the indigenous people: the right to administer the use of a variety of land in utilized to process traditional land management by applying conventional methods in their native land; their right to get financial assistance from several sources, including the Russian government (state and local), international organizations and nongovernmental organizations (NGOs); the right to receive compensation for damages caused by economic and industrial activities; and the right to protect and promote their culture (Novikova, 2016).

For their economic progress and to protect their legal rights further, Russia has recently announced new and unique legal frameworks such as the "Strategy of Development of the Arctic Zone of the Russian Federation and the Provision of National Security for the Period to 2035" and the federal law "On State Support for Entrepreneurial Activity in the Arctic Zone of the Russian Federation (2020)" (Gorbunova, 2021).

3.3 Canada

Canada captures a massive part of the Arctic. The total Arctic area under Canada's control is 40% of its territory and represents 25% of the global Arctic. The Canadian Arctic includes the Northwest Territories, Nunavut, Yukon, and the northern parts of many provinces. The Canadian Arctic covers the northern part of mainland Canada and consists of 94 big islands and 36,469 small islands. Less than 1% of Canada's population lives in this vast area. The total population of the Canadian Arctic is 200,000, and half of its population consists of indigenous people (https://www.international.gc.ca/world-monde/international_relations-relations_internationales/arctic-arctique/index.aspx?lang=eng; https://arctic-council.org/about/states/canada/; https://www.thecanadianencyclopedia.ca/en/article/arctic-archipelago; https://www.thecanadianencyclopedia.ca/en/article/arctic-archipelago).

The following are some key aspects of the management and development of the Arctic in Canada:

International Cooperation: Canada recognizes the global nature of the Arctic challenges and opportunities and collaborates with other Arctic nations on various matters. The main channel for international cooperation is the Arctic Council that was formed in Ottawa in 1996 through the Ottawa Declaration. Canada was elected as the first chair of the Arctic Council in

1996 for 2 years and appointed for the second term as a chair in 2013. Global Affairs Canada organizes and assists the international aspects of Canada's Arctic engagement. The Nordic and Polar Relations Division mainly conducts this activity in Ottawa, the Canadian International Arctic Centre (CIAC), based in Oslo, Norway, and by Canadian embassies in different countries (https://www.international.gc.ca/world-monde/international_relations-relations_internationales/arctic-arctique/index.aspx?lang=eng).

Canada joined the Ilulissat Declaration in May 2008 to resolve the boundary disputes with neighboring countries. Also, it expressed its continuous interest in collaborating with other Northern countries in the throne speech in March 2010 (Byers, 2010). Canada prefers to handle boundary conflicts in the Arctic according to international law. Canada has no quarrel with its Arctic neighbors over Canadian Arctic land except for the dispute of Hans Island with Denmark. Canada has some disputes with the United States and Denmark regarding the Beaufort and Lincoln Sea water boundary. However, these disputes have neither impacted Canada's relations with its neighbors nor resulted in any severe defense threats for the Arctic nations due to the Canadian government's excellent management of these conflicts (STRATEGY, 2010).

In 2010, in its Arctic Foreign Policy, Canada mentioned the United States as its leading collaborator in the Arctic. The United States also highlighted the importance of a unique and strong partnership between the two countries to strengthen security matters in the Arctic region (Lackenbauer & Huebert, 2014). Canada and the United States founded the North American Aerospace Defense Command (NORAD), and both countries collaborate through NORAD to better manage and administer Northern airspace. The Canadian government annually administers a sovereignty operation known as Operation Nanook in Canada's Arctic to protect and manage its control over the air, land, and sea. In 2010, Canada collaborated with the United States and Denmark in Operation Nanook to better manage cross-border threats (STRATEGY, 2010).

Northern Infrastructure Development: Investments in infrastructure are necessary for the progress of the Arctic region. Canada has been investing in enhancing transportation networks, including ports, airports, and roads, to improve connectivity and accessibility for remote communities and economic activities. According to Lamoureux et al. (2015), the strengths and protection of infrastructure in the Canadian Arctic, especially in the Yukon North Slope, Northwest Territories, and Nunavut, are central areas of interest for the local people, governments, and industrial sector.

The Government of Canada has invested in vital infrastructure to promote investment and development in the Canadian Arctic. Since 2007, Infrastructure Canada has initiated infrastructure projects of $810 million. These infrastructure development projects contribute to improving connectivity of the remote northern areas and promoting economic progress and job creation for the local people. For instance, one of the projects is to build up the Mackenzie Valley Winter Road, which has resulted in a better approach to the Northwest Territories' oil, gas, mineral resources, and hydroelectric water sources. The connectivity because of the road project also resulted in a better flow of goods and services in the area and a lower cost of living. Another exciting road project is to finish the all-season road linking Inuvik to Tuktoyaktuk. This five-year road project, which started in 2012, received $150 million from the Government of Canada (https://www.infrastructure.gc.ca/pub/north-nord-eng.html).

The following are other investment projects related to infrastructure development in the Canadian Arctic: a $71 million grant from the Government of Canada for the renovation of the Yukon Energy Corporation Mayo B Hydroelectric generation facility and execution of the Carmacks-Stewart transmission line in Yukon. The project will contribute to sustainable economic development in the region. The second project that also enhances sustainability in the region is a wastewater treatment service in Taloyoak, Nunavut, to provide clean drinking water to the people in the area (https://www.infrastructure.gc.ca/pub/north-nord-eng.html).

Indigenous Peoples' Rights: The Canadian Arctic, along with Greenland, shares the highest population of indigenous people, i.e., about 75% of the total indigenous population in the Arctic (https://nordregio.org/maps/indigenous-population-in-the-arctic/). The indigenous population in the Canadian Arctic consists of Athabaskan, Inuit, and Gwich'in tribes. The Athabaskan and Gwich'in tribes are the residents of the Northwest Territories and Yukon region of the Canadian Arctic. In contrast, the Inuit tribe resides in the northern area of Canada called Inuit Nunangat. All three indigenous tribes have permanent representations in the Arctic Council (https://arctic-council.org/about/states/canada/#:~:text=Indigenous%20Permanent%20Participants%20who%20live,the%20northern%20regions%20of%20Canada).

The indigenous people in the Canadian Arctic have faced discrimination compared to the rest of Canadian citizens in acquiring quality transportation, energy, advanced technology, job opportunities, and health and education services. Despite the Canadian governments' efforts in the past, none of their policies have resulted in sustainable economic development in the Canadian

Arctic. In 2019, the Canadian government took a bold initiative by developing the new framework "Northern Policy Framework" by involving various stakeholders such as the Federal Government, indigenous citizens, six territorial and provisional governments represented by Yukon, Northwest Territories, Nunavut, Newfoundland and Labrador, Quebec, and Manitoba. The Northern Policy Framework contains the following tasks to improve the well-being of indigenous people:

- Provide good health services to the local people and communities
- Invest in improving the energy, technology, and transport infrastructure
- Provide employment opportunities, promote innovation, and drive economic growth in the Arctic and Northern area.
- Promote scientific development, knowledge sharing, and research activities that are helpful for the local people and policymaking.
- Formulate climate change adaptation policies in the Arctic.
- Protect the safety and security of the people in the Arctic.
- Make sure that Canada remains a global Arctic leader.
- Promote close relationships and integrate indigenous people with non-indigenous residents (https://www.rcaanc-cirnac.gc.ca/eng/1560523306861/1560523330587).

Resource Extraction: The Canadian Arctic is rich in natural resources such as oil, uranium, base metals and silver, tungsten, rare earth element, iron ore, zinc, copper, diamonds, gold, and fisheries. The vital minerals contribute to renewable energy and sustainable technology (such as batteries, solar panels), advanced production, and electronics. In 2019, the worth of mineral extraction in the Canadian North was estimated as $3.3 billion, out of which Northwest Territories had a share of $1817 million, Nunavut $1315 million, and Yukon $169 million. The Canadian government encourages sustainable mining as a part of its effort to make Canada a sustainable economy in the world. Similarly, oil and gas are a source of good income for the Canadian Arctic. For instance, in 2019, oil and natural gas contributed $197 million to the Canadian Arctic's gross domestic product (GDP), resulting in 546 jobs in the oil and gas sector (https://www.cannor.gc.ca/eng/1368816364402/1368816377148).

However, it is challenging to extract these resources because of the higher costs associated with its remote location and severe climate. For instance, oil companies Chevron Corp., ExxonMobil, and BP stopped their project to explore the Arctic gas and oil drilling in the Canadian Arctic at the Beaufort Sea due to technical and operational challenges and higher extraction expenses (https://www.marinelink.com/news/canadian-drilling-arctic393698). Because the development of the Canadian Arctic and its residents mainly depends on the extraction of these natural resources, the Canadian government is taking measures, including the development of the Northern Policy Framework, to facilitate investment, create employment opportunities, protect the environment, and for the integration of indigenous people in decision-making (https://www.cannor.gc.ca/eng/1368816364402/1368816377148).

Managing and developing the Arctic in Canada requires a good balance between environmental conservation, protecting indigenous rights, responsible resource development, and cooperation with other Arctic states. The Canadian government strives to ensure the sustainable growth and well-being of the Arctic region and its inhabitants while recognizing its strategic importance internationally.

3.4 United States

In 1867, the United States bought the state of Alaska, which made it one of the Arctic nations. The Arctic territory that belongs to the United States includes the north and west of the border area made by the Porcupine, Yukon, and Kuskokwim Rivers; the contiguous seas, the Arctic Ocean, and the Beaufort, Bering, and Chukchi seas; the Aleutian chain. The development and management of the Arctic by the United States include many strategic, economic, and geopolitical factors, such as national security, sustainable development in the region, developing close relations with other Arctic countries, integrating indigenous people in policymaking, and promoting scientific research (https://arctic-council.org/about/states/the-united-states/). The following are some main areas of the management and development of the Arctic in the United States:

National Arctic Strategy: In 2013, the United States developed the National Strategy, which demonstrates the government of the United States's strategic preferences for the Arctic area. The main objective of the National Strategy is to set guidelines to deal the emerging challenges effectively and to utilize the economic opportunities that emerge from the Arctic due to the melting of the ice. The National Strategy explains the national security concerns of the U.S. government and its preference to integrate several stakeholders, such as the Federal government, local and indigenous representatives, the private sector, and international collaborators, to improve and handle the security issues. The strategy also includes measures to manage and adapt to climate changes in the Arctic region. The following are three target areas of the National Strategy: first, improve and support the U.S. security interest in the region by using ship vessels and aircraft to protect the air and sea according to international law. Second, save the Arctic environment and its natural resources and promote scientific research to understand the region better; and third, promote international cooperation through multiple sources, including the Arctic Council (Obama, 2013).

Resource Extraction: The Arctic area under U.S. control is rich in natural resources. For instance, water in Alaska carries about 27 billion barrels of oil and 132 trillion cubic feet of natural gas, and a significant portion of these resources are in the Arctic. In particular, the Alaskan Arctic's Chukchi Sea holds a substantial source of undiscovered oil and gas. However, according to Cohen (2010), despite having access to these rich resources, compared to Russia and China, the United States has yet to show much progress in resource exploration and infrastructure development in the Arctic region. This gap was mentioned in a National Petroleum Council (NPC) policy document submitted to the U.S. Secretary of Energy. Considering this deficiency, in the summer of 2015, the U.S. government permitted the exploration of resources in the Alaskan Arctic's Chukchi Sea (https://www.brookings.edu/articles/the-u-s-still-needs-arctic-energy/).

In the latest "National Strategy for the Arctic Region," which is the revised and updated form of the National Strategy developed in 2013, the U.S. government showed interest in collaborating with international and national collaborators, including the Export–Import Bank, U.S. International Development Finance Corporation, and U.S trade development Agency, to encourage private sector investment in exploring the crucial minerals. The strategy also emphasized investing in transport infrastructure to better access these natural resources (https://www.whitehouse.gov/wp-content/uploads/2022/10/National-Strategy-for-the-Arctic-Region.pdf).

Indigenous People's Rights and Development: Arctic indigenous people in the United States (Alaska) consist of the following tribes: Aleut, Alutiiq, Yup'ik, Iñupiaq (Northwest Alaskan Inuit), Athabaskan, Tlingit, and Haida. These tribes are about 18% of the total population of Alaska, and their primary source of income is fishing and hunting (https://arctic-council.org/about/states/the-united-states/). The US government keeps the policy of integrating and coordinating with Alaska's indigenous people to utilize their knowledge and involve them in critical decision-making. The latest updated National Strategy for the Arctic Region emphasized increasing the efforts for the region's sustainable development and improving the quality of life of indigenous people in Alaska. To achieve this, it is recommended to invest in infrastructure development, providing access to better quality services and supporting the business and economic sectors in the region. In 2021, the United States made a massive investment in infrastructure in transport, technology, and energy sectors to improve the country's economic condition, including Alaska. However, the government is committed to investing more in Alaska's transportation and advanced technology infrastructure to improve its economy and provide access to its indigenous people to better knowledge and education. The government is also interested in enhancing the sea and air infrastructure, like constructing a deep seaport in Nome and other small ports and airfields (https://www.whitehouse.gov/wp-content/uploads/2022/10/National-Strategy-for-the-Arctic-Region.pdf).

4 China and the Arctic

The Arctic Policy of China outlines the country's approach to foreign relations with Arctic countries as well as its plans to develop infrastructure, extend military capabilities, conduct research, and excavate resources within the Arctic Circle (Ngila, 2023). In January 2018, China has released its policy paper on the Arctic. In the paper, it draws a picture of how China perceives the economic possibilities that the region could provide. Based on such, China has vowed to actively participate in Arctic affairs as a major stakeholder—the so-called "Near-Arctic state" (Jash, 2018). In addition, China would focus on enhancing military capabilities of its army that could extend into the Arctic region.

Such an ambition has been reflected by China's plans to build new shipping routes through the Arctic based on the so-called "Polar Silk Road." This is an initiative conducted jointly by China and Russia in 2018 (Dams et al., 2020). Such an initiative was further highlighted with the Russo-Ukrainian conflict, where international sanctions from the NATO countries have forced Russia to further seek Chinese technological and financial aid to develop the Yamal-Nenets region.

In terms of Arctic shipping routes, while China has remained largely neutral on Canada's position that the Northwest Passage is within Canada's internal waters (Wright, 2011), it shows more interests on the Northern Sea Route, where the maritime shipping distance from the ports of Shanghai to Hamburg is about 4000 miles shorter through the Northern Sea Route compared to the more conventional routes via the Strait of Malacca and the Suez

Canal (Buixade Farre et al., 2014). One of the major long-term goals of China is the further opening of the Northern Sea Route (which by 2030 may be fully ice free) to shorten the shipping distance from China to Western Europe by at least 23%. Achieving such would reduce China's dependence on the conventional routes, which poses several chokepoints aligned with the United States (Dams et al., 2020). This has increasingly become critical given the importance of establishing reliable and resilient supply chains for the country. However, this is far from being straightforward, as pointed out by some experts on the Arctic on the limitations of Arctic sea routes, including harsh conditions, more icebergs due to melting of Greenland's icecap, higher insurance premiums, lack of infrastructure, and shallow depths (Buixade Farre et al., 2014).

China has remained neutral on Canada's position that the Northwest Passage is in Canada's internal waters.

5 Business and environmental management view of the Arctic

In terms of the development of the Arctic, several projects have taken place. A prominent example is the Yamal LNG projects in Russia (Afenyo et al., 2022; Zhu et al., 2023). Officially, the project is administered by the JSC Yama LNG and is a joint venture of the following companies with the corresponding percentage of stake: NOVATEK of Russia (~50%); TOTAL of France (~20%); CNPC of China (~20%); and the Silk Road Fund of China (~9.9%). The project is located at Yamal Peninsula above latitude 66 degrees north and harnesses the resources of the South Tambey Field.

These projects are generating substantial financial gains, particularly for Russia. The other Arctic countries do not directly benefit from these projects. However, collaborations with countries like France and the United States through their oil companies are the ways such benefits can be harnessed. In Canada, the government has been working on further developing the only deep-water Arctic port in Canada, notably the port of Churchill in Manitoba. Most of the Arctic countries together with their non-Arctic counter parts have engaged in scientific expeditions through different platforms like Canadian Arctic Research Expedition (see: https://seds.ca/projects/can-arx/); Beaufort Exploration Project (see: https://www.whoi.edu/beaufortgyre/history/history_modern.html); among others. At the same time, some shipping companies have tried transporting goods through the Arctic. Even though they are generally happy with the time and resources saved, many are cautious about the environmental implications. Therefore, in 2019, through the Arctic Corporate Shipping pledge, a lot of companies decided to abandon the idea of shipping through the Arctic, such as Nike, Ralph Lauren Corporation, Kuehne+Nagel, PUMA, International Direct Packaging, Allbirds, Aritzia, Hudson Shipping Lines, Bureo, to name but a few. Also, there is continuous exploration and development and production of oil and gas in this region. Environmentalists, however, are of the opinion that the Arctic should be left alone, and the flora and fauna preserved. This means that all development and business-related activities in the Arctic need to stop. However, opening the Arctic more to business would enable the indigenous communities living in this area have access to cheap and timely supplies for food and medicine. The question of exploring the Arctic for nonrenewable resources and shipping through the region largely lies with the countries involved. While a country like Russia is aggressive in its development of the Arctic, on the contrary, Canada is taking a more conservation approach. In this case, Russia has developed a special division in their Armed forces mainly focused on the Arctic. This is no surprise as 53% of the Arctic coastline belongs to Russia.

6 Conclusions

The Arctic is one of the most important prospects for development and business. Even though there are opportunities in this area, there is also environmental challenges. In terms of transportation, supply chain and shipping companies would benefit much by using the Arctic routes, including substantial savings on money, time, and resources. However, the environmental implications create uneasiness in the business community. Indeed, the potential for oil spills during oil and gas activities further makes things complicated. Should an oil spill occur in the Arctic, it could pollute even flora and fauna under the ice and affect the livelihood of the indigenous people. Since the Arctic Council now is virtually just a work in progress that does not really have a legal framework for enforcement of the recommendations, in the foreseeable future, it would be very challenging to get business done in this region in an absolutely safe and sustainable way. It would be extremely important to first empower the Arctic Council before such steps are taken. Now enforcing Arctic policy recommendations rests on the individual Arctic countries. Different countries have their policies driven by national interest and so, while some may well be opened for business, others may not. Corporate organizations have in most parts pledged to refrain from shipping goods through the Arctic. The blockage of the Suez Canal by the container vessel *Ever Given* in March 2021, however, might have just changed this way of thinking. This is because the Arctic could realistically serve as an alternative route for the movement of the goods across the world and avoid the dangers of getting locked up in the canal, which have proved to lead to a significant loss of the world's GDP in the past years.

References

Afenyo, M., Jiang, C., Ng, A. K. Y., & Lee, P. T. (2022). A Bayesian-loss function-based method in assessing loss caused by ship-source oil spills in the arctic area. *Risk analysis: An Official publication of the Society for Risk Analysis, 48*, 8. https://doi.org/10.1111/risa.14025.

Afenyo, M., Ng, A. K. Y., & Jiang, C. (2021). A multi-period model for assessing the socio-economic impact of oil spills in the Arctic. *Risk Analysis: An international Journal., 42*(3), 614–633.

Bortnikov, N. S., Lobanov, K. V., Volkov, A. V., Galyamov, A. L., Vikent'Ev, I. V., Tarasov, N. N., … Chefranov, R. M. (2015). Strategic metal deposits of the Arctic Zone. *Geology of Ore Deposits, 57*, 433–453.

Buixade Farre, A., Stephenson, S. R., Chen, L., Czub, M., Dai, Y., Demchev, D., Efimov, Y., Graczyk, P., Grythe, H., Keil, K., Kivekäs, N., Kumar, N., Liu, N., Matelenok, I., Myksvoll, M., O'Leary, D., Olsen, J., Pavithran, A. P. S., Petersen, E., … Wighting, J. (2014). Commercial Arctic shipping through the northeast passage: Routes, resources, governance, technology, and infrastructure. *Polar Geography, 37*(4), 298–324.

Byers, M. (2010). Cold peace: Arctic cooperation and Canadian foreign policy. *International Journal, 65*(4), 899–912.

Cohen, A. (2010). *From Russian competition to natural resources access: Recasting US Arctic policy*. Heritage Foundation.

Dams, T., van Schaik, L., & Stoetman, A. (2020). *Presence before power: Why China became a near-Arctic state (report)* (pp. 6–19). Clingendael Institute.

De Agreda, A. G. (2017). From cylinder to sphere: The impact of climate change in the Arctic beyond the polar circle. In E. Conde, & S. I. Sanchez (Eds.), *Global challenges in the Arctic region: Sovereignty, environment and geopolitical balance* (pp. 347–363). London and New York: Routledge.

Dodds, K., & Nuttall, M. (2019). *The arctic: What everyone needs to know®*. Incorporated: Oxford University Press.

Frolov, I. E. (2015). Development of the Russian Arctic zone: Challenges facing the renovation of transport and military infrastructure. *Studies on Russian Economic Development, 26*, 561–566.

Gorbunova, A. (2021). Legal consideration regarding indigenous people of the Russian Arctic, polar. *Perspectives, No.*7, 1–14.

Gritsenko, D., & Efimova, E. (2020). Is there Arctic resource curse? Evidence from the Russian Arctic regions. *Resources Policy, 65*, 101547.

High North News. (2023). *Italy's growing interest in the Arctic.* Accessible at https://www.highnorthnews.com/en/italys-growing-interest-arctic.

Hønneland, G. (2006). *Kvotekamp og kystsolidaritet. Norskrussisk fiskeriforvaltning gjennom 30 ar°*. Bergen: Fagbokforlaget.

Jash, A. (2018). *IPP REVIEW—China's need to build the "polar silk road"*. Archived from the original on 2018-09-04. Retrieved from Ippreview.com. (Accessed 4 September 2018).

Jensen, Ø., & Rottem, S. V. (2010). The politics of security and international law in Norway's Arctic waters. *Polar Record, 46*(1), 75–83.

Kovalev, A. A. (2004). *Contemporary issues of the law of the sea: Modern Russian approaches. Vol. 1.* Eleven International Publishing.

Lackenbauer, W., & Huebert, R. (2014). Premier partners: Canada, the United States and Arctic security. *Canadian Foreign Policy Journal, 20*(3), 320–333.

Lamoureux, S., Forbes, D. L., Bell, T., Manson, G. K., Rudy, A. C. A., Lalonde, J., et al. (2015). The impact of climate change on infrastructure in the western and central Canadian Arctic. In G. A. Stern, & A. Gaden (Eds.), *From Science to Policy in the Western and Central Canadian Arctic: An Integrated Regional Impact Study (IRIS) of climate change and modernization* (pp. 301–341). ArcticNet.

Ngila, F. (2023). *China is cementing its position as an Arctic superpower through Russia.* Quartz. Retrieved 20 April 2023.

North, B. I., & Bullvåg, E. (2017). *A periodic report with insight to business activity and opportunities in the Arctic.*

Norwegian Ministry of Foreign Affairs. (2021). *The Norwegian Government's Arctic Policy People, opportunities and Norwegian interests in the Arctic.* Available to download at https://www.regjeringen.no/en/dokumenter/arctic_policy/id2830120.

Novikova, N. I. (2016). Who is responsible for the Russian Arctic?: Co-operation between indigenous peoples and industrial companies in the context of legal pluralism. *Energy Research & Social Science, 16*, 98–110.

Obama, B. (2013). *National strategy for the Arctic region.*

Østhagen, A., Levi Sharp, G., & Sigurd Hilde, P. (2018). At opposite poles: Canada's and Norway's approaches to security in the Arctic. *The Polar Journal, 8*(1), 163–181.

Rottem, S. V. (2019). *The arctic council: Between environmental protection and geopolitics.* Springer Singapore Pte. Limited.

Serova, N. A., & Serova, V. A. (2019). Critical tendencies of the transport infrastructure development in the Russian Arctic. *Social and Economic Development, 36*, 35.

Shaw, M. N. (2017). *International law.* Cambridge University Press.

Snetkov, V. N., Weshkelsky, A. S., & Yurkova, O. Y. (2020, July). Historical-legal aspect of the political system in the Russia's arctic region. In *Vol. 539. IOP conference series: Earth and environmental science* (p. 012009). IOP Publishing.

Stoltenberg, J. (2013). Managing the Arctic: A Norwegian Perspective. *Harvard International Review.* Available to download at https://hir.harvard.edu/managing-the-arctic-a-norwegian-perspective/.

STRATEGY, N. (2010). Statement on Canada's Arctic foreign policy. *Lokaliseret d, 28*(5), 2012.

Underhill, W. (2007). Norwegian gold. *Newsweek.* 8 January 2007 http://www.newsweek.com/id/56729. (Accessed 13 June 2023).

USDS, (n.d.). Office of Polar and Ocean Affairs. Available at https://www.state.gov/key-topics-office-of-ocean-and-polar-affairs/arctic/#:~:text=The%20geographic%20area%2C%20as%20defined,adjacent%20marine%20areas%20in%20the Assessed 24 May 2023.

Wright, D. C. (2011). *The dragon eyes the top of the world: Arctic policy debate and discussion in China.* Naval War College, China Maritime Studies Institute (No. 8).

Zhu, S., Ng, A. K. Y., Afenyo, M., Panahi, R., & Bell, M. G. H. (2023). Socio-economic impacts of shipping along the Northwest Passage: The cost to locals. *Marine Policy, 153*, 105647.

CHAPTER THREE

Arctic policies, governance, and cooperation: A catalyst for the development and management of the Arctic

1 Introduction

As alluded to in the previous chapter, the policies are based very much on individual countries' interest. However, recommendations do emanate from the Arctic council meetings and various working groups. These recommendations cannot be implemented by the Arctic council as they do not have the legal authority to do so. In this chapter, we examine some of the individual policies of some of the countries in the Arctic as well as near-Arctic countries. A few countries were chosen for the purpose of discussion in this chapter. Many countries have already published policies regarding their plans for development and research including the United States (US), Canada, Russia, India, and Norway. Each of these policies has unique strengths and weakness that cater to each individually.

2 Selected Arctic countries policy

The U.S. Arctic policy contains four pillars and is largely focused on cooperation. These pillars include infrastructure development, environmental protection, sustainable economic development, and international cooperation. The policy interestingly goes out of its way to specifically mention the US allies and address the ongoing Russo–Ukrainian conflict. The United States has taken a unique approach and has clarified it will neither assist Russia with Arctic development nor share its resources if the current Ukraine conflict continues. The details of the pillars as grouped in terms the themes are discussed. For security, the United States seeks to develop capabilities to expand Arctic activities (The White House (TWH), 2022). This

Safe and Sustainable Arctic Shipping Management and Development
https://doi.org/10.1016/B978-0-443-18849-7.00011-7
Copyright © 2024 Elsevier Inc.
All rights are reserved, including those for text and data mining, AI training, and similar technologies.

will entail the capabilities required to defend her interest in the Arctic while coordinating shared approaches. For climate change and environmental protection, the United States pursues international initiatives to mitigate emissions in the Artic (TWH, 2022). Sustainable Economic Development, on the other hand, increases responsible Arctic investment, and finally, through International Cooperation and Governance, the United States works to sustain institutions for Arctic cooperation (TWH, 2022).

Canada has a very different approach to the Arctic. Much like the United States, it mentions infrastructure and economic development. However, it specifically places emphasis on the lifestyle of the Canadian citizens in the Arctic region (Government of Canada (GC), 2019). Canada's vision is to create a strong, self-reliant, prosperous community (GC, 2019). The indigenous population of the Inuit have largely been lacking infrastructure and cultural equity compared to Canadians in more populous areas (GC, 2019). The actions with regards to the Canadian Arctic policies are that it is extremely important to invest in energy, transportation, as well as infrastructure related to communication that are very essential to the survival of northern and Arctic governments, economies as well as the respective communities (GC, 2019). In addition, the Canadian Arctic policy emphasizes the need for investment in science through the support of knowledge and research that is meaningful for communities and for decision-making. There is also the critical need of supporting healthy ecosystems in the Arctic and the North as the effects of climate change are only getting worse. The security of Canadians is very critical, and so the Arctic policy provides a system to ensure that Canada and our Northern and Artic residents are safe, secure, and well defended (GC, 2019). International relations are very critical to Canadian corporation on solving Arctic issues and so the Canadian Arctic policy further seeks to restore Canada's place as an international Artic leader (GC, 2019). The bulk of the Arctic is inhabited by the indigenous people spreading over more than 400 communities, and therefore, the policy also promotes reconciliation and improve relationship between indigenous and non-indigenous people (GC, 2019).

Unlike the previous two policies, the Russian Arctic policy takes a highly economic and strategic approach. At the time of writing this chapter, the Russian Federation has been suspended from the Arctic Council mainly due to its involvement in the Ukrainian conflict. Russia has a long northern sea border that will be crucial for expanding transportation, trade, and military capabilities. This new access to Arctic resources will mean an increase in Russian economy by gross domestic product (GDP). Russia specifically mentions growth in the oil and gas sectors with plans to expand and build liquefied natural gas (LNG) terminals. Effectively, they hope to turn into a new energy superpower in the market. In addition to this, they have identified the United States as their primary geopolitical competitor and China as a threat to its resources and foothold in the Arctic. This is not surprising given the Arctic policy of the United States. The Russian Arctic policy therefore entails the following:

(i) Developing untapped energy reservoirs to elevate its economy. Russia is already producing oil and gas from the Arctic and plans to expand further through collaboration with other countries (Ahmad & Zafar, 2022).
(ii) Utilization of the Arctic as strategic resource base to fulfill the country's socioeconomic needs (Ahmad & Zafar, 2022).
(iii) Look for alternative relationships in the Artic (Ahmad & Zafar, 2022).
(iv) Reassert its prestige and Power (Ahmad & Zafar, 2022).
(v) Secure transport routes and prepare for conflict (Ahmad & Zafar, 2022).
(vi) Develop cities and economies in the far North (Ahmad & Zafar, 2022).

Norway's policy is a lot like the United States. Their goals also focus on peace, stability, and value creation. Also, it mentions Russia and China as threats to the region and other countries. A large section of the policy revolves around maintaining international cooperation and enforcement of international legal order (Norwegian Ministry of Foreign Affairs (NMFF), 2020). It suggests using the Law of the Sea as an international framework for monitoring activities in the Arctic Ocean as well as using the North Atlantic Treaty Organization (NATO) and the United States Coast Guard (USCG) for the area's security (NMFF, 2020).

The key themes of Norwegian Arctic policy are shown in Fig. 1.

In terms of Peace, Stability, and Predictability, Norwegians seek to have strong and credible cooperation when it comes to defense with their allies as well as other relevant partners to make this theme a reality.

By the second theme, which is the international cooperation and the international legal order, the Norwegian government would endeavor to present only facts during international debates about the Arctic. They hope to do by engaging in rigorous research exercises. The other theme is integrated, ecosystem-based management, which is geared toward transforming the economy into a society that takes in extremely low carbon emissions and by extension becomes a low carbon society. The increase in job creation and value creation is another critical theme and would be achieved through the support and development of new, environmentally friendly cities in the region. Since most of the population in this region requires a lot of help from the central government to bridge the gap with the others in the country, the final theme is having Effective Welfare schemes. This is created so that the

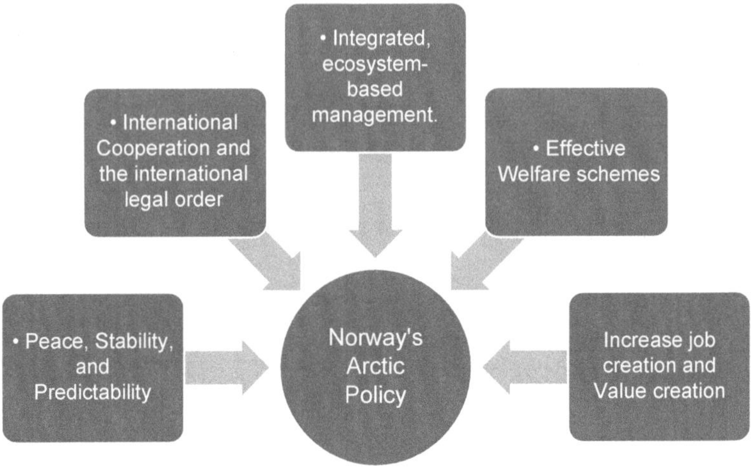

Fig. 1 The pillars of the Norway's Arctic policy.

government can preserve the identity and culture of national indigenous communities and to empower them to become self-sufficient and economic independence (NMFF, 2020).

3 Selected non-Arctic countries policy

India's Arctic policy is different from all those mentioned before. Their goals are to increase their presence and cooperation with the Arctic region, gain polar research knowledge, and increase scientific development and research in the area. India is unique because its main goal is R&D specifically concerning the Himalayas. They refer to them as the "third pole" and believe studying a similar landscape would assist in research efforts (Government of India (GI), 2022). Because they have already highly researched the Himalayas, they argue that they can contribute decades of polar research to the international scientific community. One notable and interesting point is India's concern with the thawing of the permafrost and the release of dormant viruses and bacteria which could present a global threat (GI, 2022). India's policy is largely based on scientific cooperation and less on the economic and political aspects.

The pillars of the Indian Arctic Policy can be found in Fig. 2. As part of the Science Research pillar, India seeks to encourage research to align with international Arctic Policies. In addition, through the Climate and Environmental Protection pillar, they seek to enhance the quality of India's engagement with the Artic Region. One of the other pillars is Economic and

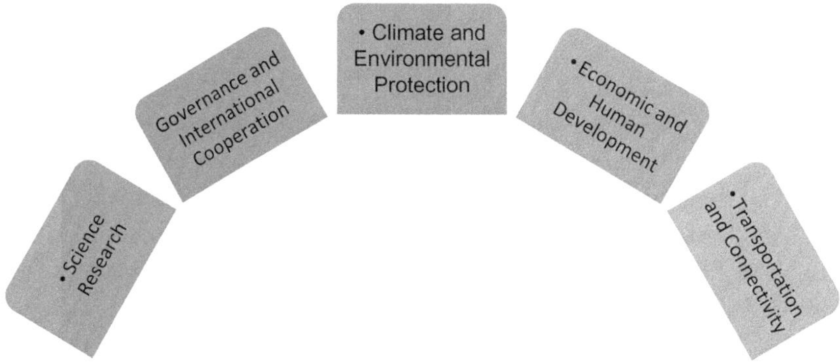

Fig. 2 The pillars of the Indian Arctic policy.

Human Development. India proposes to do this in their policy by pursuing collaboration with Arctic States, for mutual beneficial and sustainable economic cooperation and investment. In addition, they will provide healthcare services and technological solutions to indigenous people. Transportation and connectivity is another important aspect of the India's Arctic policy. The suggestion in the policy is that they intend to collaborate with shipbuilders who have the knowledge of making better suited and sustainable ships equipped for the Arctic. The improvement of communication infrastructure is another critical step the country intends to take to address this pillar (GI, 2022).

Governance and international cooperation is very important as no country is an island or an isolation. India seeks to promote security and stability in accordance with international treaties and covenants (GI, 2022).

China, which is a near-Arctic country, has become a very important player on the world stage both in terms of security and economy. The country has identified the Polar Silk Road as one of the main pillars of its Belt and Road Initiative (BRI). They have collaborated with Russia on the LNG projects in the Arctic. China's Arctic policy is summarized in the following sentences:

1. Understand, protect, develop, and participate in the governance of the Arctic to make sure the common interests of all the countries are met (State Council Information Office of China (SCIOC), 2018).
2. Prioritizes scientific research and commits itself to maintain a peaceful, secure, and stable Arctic order (SCIOC, 2018).
3. Ensuring access to commercial opportunities (SCIOC, 2018).
4. Build capabilities to enforce its perceived rights and claims to the region (SCIOC, 2018).
5. Promoting peace and stability in the Arctic (SCIOC, 2018).

Mariia (2019) explained that China's Arctic policy is based on three internal features: first, Arctic policy is established on the national interests such as the BRI, gaining economic power and target of playing the role of a significantly essential and responsible country in the world. These national interests are reflected in China's action of integrating Arctic activities into the state agenda. For instance, the Polar region was regarded as the National Security Law of China, and consequently, it emphasized China's potential contribution to rescue and military activities abroad. China's 13th five-year plan also included the development of a new icebreaker and stressed the need for cooperation which should be facilitated by new observation basecamps in the Arctic. In 2017, China integrated the Arctic into its BRI strategy and proposed international cooperation in expanding the BRI activities to the Arctic, such as the potential construction of the "Arctic Silk Road." Second, China consulted specialists and scholars in the conceptual formulation of the Arctic policy. The main objective of involving the scholars was to effectively express China as a dominant country and potential emerging power and its role in the Arctic. China is in favor of the big Arctic, which consists of not only the Arctic nations but also non-Arctic countries, not only those who already showed interest in their involvement in the Arctic, but also new interested nations such as Mongolia, Kazakhstan, Uzbekistan, Turkmenistan, Tajikistan, and Afghanistan. From a military point of view, the Arctic is strategically important for China, and due to the decreased political interest of the United States in the matters of the Nordic countries and a potential increase in the military power of Russia, China is perceived as a new source of the security in the region. Another unique characteristic of China's Arctic policy is that each province or large city has the autonomy to play a role in the Arctic as a part of China's Arctic policy. For example, the Heilongjiang Province' is collaborating with two Russian cities, Murmansk and Arkhangelsk, for Arctic development. In contrast, Shanghai focuses on developing collaboration with Arctic Nations in the maritime sector. In conclusion, China's Arctic policy is based on all three features that are national interest, experts' involvement, and regionally specific.

4 Policy similarities

Many of these policies are similar. Almost all policies include sections regarding the economic, political, social, and environmental aspects of Arctic development. Both the United States and Norway have very similar goals regarding the protection of their nations and allies from threats such as Russia and China. Almost all policies except Russia have put large emphasis on environmental protections and economic sustainability in the event of a crisis.

Many of these policies have differences too. Most such differences come in the form of a larger more cohesive difference in the policy overall. For example, Canada has a large social focus and places emphasis on infrastructure development for the people and the inclusion of indigenous community. Russia has a commercial emphasis and heavily expands on Northern Sea Route (NSR) trading routes and increasing their economy. Russian policy has a distinct lack of social and environmental concerns. India has a heavily scientific approach and encourages international research cooperation and environmental protection. On the other hand, the US policy has many strengths and weaknesses. It has a clear response to the threats Russia and China present and addresses them accordingly. Additionally, it defines a response to managing and mitigating emissions in the region that no other policy does as thoroughly. Finally, it addresses native Alaskans and ensures cooperation with the state government.

All the above being said, the United States was one of the last countries on the list to publish their Arctic policy. Many other countries published their months or years before the United States who published theirs in October of 2022. The policy itself is vague in detailing expansion plans or any current infrastructure concerning ice breaking equipment and previously defined trade routes. Other countries such as Russia have a more detailed plan to expand economic growth. Finally, the United States puts more focus on the international side of its policy rather than how the United States itself can benefit from the region. India and Russia both did this by including the NSR and the Himalayas in their policies which are both region-specific bonuses from Arctic expansion.

To combat these weaknesses, the United States must keep on top of world affairs and new developments regarding the Arctic. Publishing information and policy regarding new developments in a timely format will assure the public that it has a plan and will serve to dissuade other countries and secure its foothold in the region. In addition, more detailed plans should be defined to start the development and construction of infrastructure in the region. Due to Alaska being isolated from other states, this presents a potentially challenging logistical issue if a crisis arises.

Overall, development in the Arctic region is inevitable, and more information will be gathered and published by various countries. Each country offers a unique and largely complex solution to the challenges the region presents. Because the region is rich in natural resources, the competition will continue to remain intense.

The Arctic countries have also operated on the following principles that runs through all the Arctic policies of the various countries. These are partnerships, security, and research on climate change. Partnerships are very key to all the Arctic policies of the various countries. For example, due to Russia's military actions in Ukraine, NATO unity has doubled its partnerships with the Arctic. Moreover, the United States plans to initiate partnerships based on the security of the nation. This would include Canada, Norway, Sweden, Iceland, Finland, and Denmark. Norway also plans to partner with the United States and allied nations in relation to the Arctic for the security of their nation. India plans to collaborate with the Arctic states that are mutually beneficial for science, the environment, and the economy of their country. Canada, on the other side, stated they would be working with the United Nations and the International Maritime Organization with the purpose of helping Canadians develop economically, socially, and environmentally in the rural areas of the Arctic.

Security is another theme that runs throughout the Arctic policies of the various countries. It is concerning that the main risks from the rapid melting of the ice in the Arctic have not been properly documented and incorporated into the deliberations of the Arctic council. Norway wants to get NATO to focus on transatlantic reinforcements and the importance of allied security. The main concern for the United States is the security of the nation and the increased interest Russia has in the Arctic area is the reason the United States is increasing security in the area. Canada also wants to increase their Canadian Coast Guard for the Arctic. Emergencies for the local people and emergency threats created by maritime transportation have been something the Canadian government has been working on for some time now. While China does not seem to have an interest in a military conflict with the United States in the Arctic, Russia intends to increase its military presence to deny the United States its stake in the Arctic.

On the issue of the climate change research, Norway has invested tremendously in fisheries, and other natural resources research. Near-Arctic countries like China has many science researchers in Greenland and on science vessels studying the Arctic and the impact of climate change. Major research areas include how the melting ice would affect the migration of fish and birds as well as the implications on public policy. In addition, in 2023, India plans to launch a satellite to observe the ecosystem of the Arctic.

Another area of priority is infrastructure development in marine transportation. All the countries were very interested in the resources located in the Arctic waters, for example, oil and gas, renewable energy, and minerals. There is also emphasis on promoting job opportunities for the people who live in these regions. Also of importance is the preservation of the culture of the people. Some of the Arctic countries have also prioritized expanding technology services to remote areas. The technologies can be applied to remote education and telehealth. To address the economic challenges, the governments of the various Arctic countries are encouraging businesses to move north. Canada particularly highlighted the need to assist the current community in the Arctic.

Norway, among other things, wants to increase its dropping population in the Arctic by creating more job opportunities. They have developed projects in the area to attract visitors and guests alike to attract young people and businesses. India wishes to make the northern Atlantic into a tourist area to help locals with job creation.

5 The Polar Code

On January 1, 2017, the International Maritime Organization (IMO) introduced the International Code for Ships Operating in Polar Waters (Polar Code). The primary purposes of developing the Polar Code were to safeguard the ships' operations in polar waters, including the Arctic and the Antarctic, and to take extra measures to prevent the polar environment by taking into account the factors and strategies that were not adequately considered and implemented by other previous regulations of the IMO (Engtrø et al., 2020).

The Polar Code considers several operational and environmental challenges related to the Arctic region. The following are some of the challenges addressed by the Polar Code:

1. **Ship Design and Equipment:** The Polar Code requires specific criteria for designing and manufacturing ships planned to operate in the Arctic water. The manufacturing guideline provides details related to the necessary features of the ship, like strong hulls, better quality in terms of stability, and an excellent machinery system to handle extreme weather conditions and operation in icy water. The Code made it mandatory for ships operating in polar waters to acquire a Polar Ship Certificate. The certificate distributes the ships into the following three groups: (a) the first group is called "Category A ship," which has features capacity to operate in medium first-year ice; (b) the second category of ships is manufactured to work in at least thin first-year ice; the last category of

vessels is specialized to operate in icy water, which is less severe than the previous two categories.

2. **Crew Training Requirements:** The code mandates that crew members, including masters, chief mates, and officers in charge of a navigational monitor on polar-operating ships, receive specialized training and have the required skills to operate in icy conditions, handle the emergencies, and adequately manage the risks due to polar environments. The training procedure must consider the framework of the International Convention on the Standard of Training, Certification, and Watchkeeping for Seafarers (STCW) and its associated law.
3. **Operational Guidelines:** The Code made it mandatory for the staff of the vessels, including the owner, operator, master, and crew, to read and follow a Polar Water Operational Manual. The manual helps provide information about the operational capacity and weaknesses of the vessel. This information is helpful for the staff to make a good decision in case of an emergency. The manual also provides optimal route planning, ice navigation, and communication guidelines.
4. **Environmental Protection:** One of the most essential features of the Polar Code is its focus on mitigating the negative influence on the environment due to vessel operations in the Arctic. The Code includes goals and requirements to prevent oil pollution, sewage management, and waste management to protect the environment during voyages in Polar water (https://www.imo.org/en/ourwork/safety/pages/polar-code.aspx#:~:text=The%20Polar%20Code%20covers%20the,waters%20surrounding%20the%20two%20poles).

6 The Barents Euro-Arctic transport plan

The following Arctic countries form the Barents region, also known as the Euro-Arctic Region: the northernmost area of Norway, Sweden, Finland, and North-West Russia. The area under the Barents region is 1.75 million km^2, and 75% of the total area is under Russian jurisdiction. The region's population is over 5 million, including indigenous people. The Sami live in the whole Barent region, while two tribes, which are Nenets and Veps, reside in the Russian area (https://barents-council.org/about-us/the-barents-region).

The industrial sectors of the four countries in the region constitute a significant portion of the export volume of each of these countries. The road is the most dominant mode of transportation for people and cargo within the region and across the border. The rail and sea modes, which have high capacity compared to road, are also utilized to export products, mainly raw materials from the region. Thus, an excellent transport system is critical for the region's economic growth and the potential development of essential industries. All four countries in the Barent region agree to establish an efficient and sustainable transport network to improve connectivity among the countries within the Barent area and with the global market. The policymakers recommend that the United Nation's sustainable development goals, including road safety, should be considered in establishing the transport network in the region. The region has good potential for the growth of the tourism industry because of the attractive tourist destinations in the area. Therefore, there is a need to improve public transport in the region. Considering the need to improve transport infrastructure and connectivity, the Joint Barents Transport Plan recommends the following transport development policies:

1. Improve the information and awareness about the region's transport network and requirements.
2. Guide and facilitate the users of the transport system to mitigate greenhouse gas emissions.
3. Provide safety on road and sea.
4. Improve the efficiency and flow of the transport system in the region.

These strategies are grouped as having short-term, mid-term, and long-term applications (https://www.vegvesen.no/globalassets/nyheter/jbtp2019-main-report-190910.pdf).

7 Discussions and conclusions

The Arctic and near-Arctic countries have different policies for which they have developed to govern how their Arctic region should be managed. Even though there are differences, there are also similarities. These similarities should be a cornerstone for the development of rigorous inter-Arctic regulations to speed up the development of the region. Furthermore, these policies should be synchronized to produce a united policy that would be representative of all Arctic and non-Arctic countries bearing in mind that currently each country is responsible for enforcing contents of the policy. A United Nation's style organization needs to be developed for the Arctic as well. Currently, the Arctic council does not have adequate powers to implement the contents of the Arctic policies of the individual countries as well as suggestions at the council meetings. This will be critical to developing the prospects of the Arctic and governing it in a peaceful manner. Already the Russia-Ukraine war exposed some of the weaknesses of the council through the expulsion of the Russian Federation. This implies that Russia may no longer collaborate on research as well as any emergency response efforts should there be need for it. While an instrument like the Polar Code is a good example, it has its own challenges and is the focus of another chapter. Environmentalists remain skeptical about how powerful such an instrument is to protect the pristine Arctic environment. Issues of carbon emissions from the fuels that are used by most ships around the world is another. The issue of the indigenous settlers in the Arctic is so important that they have special groups that represent them at the Arctic council. Many questions still linger. Will these groups be working independent of their national government at the council or will the national government policy on the Arctic supersede theirs? No matter what the answer may be, it is extremely important that the various indigenous groups are engaged and involved both at the local, national, and international level on the development and the management of the Arctic.

References

Ahmad, S., & Zafar, M. A. (2022). *Russia's reimagined Arctic in the age of geopolitical competition.* Available at https://www.airuniversity.af.edu/JIPA/Display/Article/2959221/russias-reimagined-arctic-in-the-age-of-geopolitical-competition/#:~:text=Russia%20Arctic%20Policy%202020%E2%80%932035,for%20Russia%20in%20the%20Arctic. (Accessed 31 May 2023).

Engtrø, E., Gudmestad, O. T., & Njå, O. (2020). Implementation of the polar code. *Arctic Review on Law and Politics*, *11*, 47–69.

Government of Canada (GC). (2019). *Canada's Arctic and Northern Policy framework.* Available at https://www.rcaanc-cirnac.gc.ca/eng/1560523306861/1560523330587. (Accessed 31 May 2023).

Government of India (GI). (2022). *India's Arctic Policy. Building a partnership for sustainable development.* Available at https://www.moes.gov.in/sites/default/files/2022-03/compressed-SINGLE-PAGE-ENGLISH.pdf. (Accessed 31 May 2023).

Mariia, K. (2019). China's Arctic policy: Present and future. *The Polar Journal*, *9*(1), 94–112.

Norwegian Ministry of Foreign Affairs (NMFF). (2020). *The Norwegian Government's Arctic Policy: People, opportunities, and Norwegian interests in the Arctic.* Available at https://www.regjeringen.no/globalassets/departementene/ud/vedlegg/nord/arctic_strategy.pdf. (Accessed 31 May 2023).

State Council Information Office of China (SCIOC). (2018). *China's Arctic policy. White paper.* Available at http://english.scio.gov.cn/node_8002680.html#:~:text=China's%20policy%20goals%20on%20the,sustainable%20development%20of%20the%20Arctic. (Accessed 31 May 2023).

The White House (TWH). (2022). *National strategy for Artic region.* Available at https://www.whitehouse.gov/wp-content/uploads/2022/10/National-Strategy-for-the-Arctic-Region.pdf. (Accessed 31 May 2023).

SECTION B

Economic opportunities and risks

CHAPTER FOUR

Potential consequences of Arctic shipping and resource exploitation

1 Introduction

The melting of ice has paved the way for the opportunity to extract natural resources. However, this opportunity also comes with risks (Afenyo et al., 2019, 2021, 2022). These risks can be viewed in different ways and include the risk posed by Arctic shipping and resource exploitation on the environment, the economy, the social dynamics of the region, and the governance and laws within the framework of the countries involved (Afenyo et al., 2022). In terms of the environment, the potential for an oil spill to occur is massive and the consequences to follow are dire. This is because when an oil spills in open water, cleaning it is very difficult and expensive, and when it occurs in ice-covered waters it poses more challenges to the authorities. This is because the oil could get under the ice; it could become encapsulated in ice. In addition, the oil content is also toxic to marine fauna and flora (Afenyo et al., 2016).

Another consequence is the release of black carbon into the atmosphere from ships, which will cause an increase in global warming (Browse et al., 2012, 2013). Furthermore, the ones that fall on ice help accelerate melting of the ice. This development will further add to the quantity of carbon-related emissions in the atmosphere and the level of pollution in the region (Browse et al., 2013).

Each of the Arctic councils has its own experiences with how they are been impacted by shipping taking place in the region as well as the accompanying natural resource exploration. It should be noted that the Arctic does not only contain oil and gas, as well as fauna and flora, but also precious minerals like diamonds (Afenyo et al., 2019, 2021). The technology that would be used to extract some of these minerals is a topic for another book.

Safe and Sustainable Arctic Shipping Management and Development
http://doi.org/10.1016/B978-0-443-18849-7.00012-9
Copyright © 2024 Elsevier Inc.
All rights are reserved, including those for text and data mining, AI training, and similar technologies.

The next section will examine the current impact that such activities are having on the environment, social makeup, and the economies of selected countries.

1.1 Consequences in Russia

The Russian Arctic has a considerable share of natural resources; for instance, it has 26 fields with 0.6 billion tons of oil and 8.5 trillion cubic meters of gas. To explore these natural resources, the Russian government has planned to invest 5 trillion rubles to start 150 projects by 2050. Several stakeholders, including the government, private sectors, and state organizations, have been actively involved in resource extraction in Russia (Romasheva & Dmitrieva, 2021). Thus, Russia has aggressively been developing Arctic resources in its region. In this light, the government has resorted to a tax rebate scheme to attract foreign investors to get the necessary funding for it. For example, the famous Yamal LNG projects have been successful mainly due to the collaborative efforts by different companies namely from France, China, and the United States to some extent (Afenyo et al., 2021). The project involves integrating the Sabetta seaport, a new airport, and a power plant. Some of the potential impacts of the Arctic shipping and resource exploitation in Russia are the following:

Oil and gas companies are a significant source to cause damage to the ecological balance and environment in the Arctic (Lifshits et al., 2021; Litvinenko et al., 2020). Oil and gas exploration and production in the Arctic region is more complex than the other regions due to large drilling volumes, handling huge cargo, bunkering operations, and difficult geological and severe weather conditions. During the execution of oil and gas projects, ocean resources, the earth's surface, and the environment face a massive technogenic weight, thus harming the environment. The Arctic ecosystem, which consists of various flora and fauna, is negatively affected by environmental changes. Further oil and gas activities can mitigate the sustainable development of the Arctic ecosystem, which already, due to fragility and distinctiveness, has a low probability of sustainability (Romasheva & Dmitrieva, 2021).

Resource exploitation in the Russian Arctic can generate substantial economic benefits, including profit from exporting natural resources and fisheries. Russia is number three in the world regarding natural resources, and its major export destinations are Europe and former Soviet states. However, volatility in the prices of natural resources and restrictions on Russia due to the Ukraine war has made the contribution of these natural resources to Russian economic growth uncertain (Korkmaz, 2022). The Ukraine war has resulted in a drastic decline in Russian energy prices and reduced the potential economic growth in the short and long terms (Li et al., 2023). The primary source of revenue for Russia is rents generated by natural resources; for example, Russia earns $900 million daily on average by supplying natural gas to Europe. The reduction in energy prices could have the following threefold impacts on the Russian economy: firstly, a decline in economic growth; secondly, a reduction in human and financial progress; and thirdly, a fall in employment, real wages, and a rise in inflation (Goldstein et al., 2022; Yildirim et al., 2020). Resource exploration activities have a grave negative effect on the surroundings and well-being of Indigenous people, such as the deterioration of land and sea due to oil spills, damage to the greenery of the land, and fragmentation of the area utilized for reindeer herding and hunting (Tulaeva et al., 2019).

1.2 Consequences in Greenland and Denmark

Geographically, The Kingdom of Denmark has a central location in the Arctic. The Kingdom consists of Denmark, Greenland, and the Faroe Islands. The former two have self-government. The governments of all three areas of the Realm have rights and responsibility for the resources in the Arctic (http://library.arcticportal.org/1890/1/DENMARK.pdf). Greenland is endowed with iron ore, lead, zinc, diamonds, gold, rare earth elements, uranium, and oil. This rich resource has presented an opportunity for the government to develop these and make its economy self-sustaining and eventually become independent from Denmark. In 2021, a report on the environment and nature of the DW outlet highlighted the approach adopted by the Greenlandic government. They have evaluated the prospect of oil and gas extraction on the island and came to the conclusion that the risks are too high compared to the potential benefits (see https://www.dw.com/en/greenland-stops-oil-and-gas-exploration-climate-costs-too-high/a-58294024) (Hanaček et al., 2022) and so have decided to abandon developing their Arctic resources. Natural resource exploitation, increased shipping, and traffic in Greenland may negatively impact biodiversity, habitats, and ecosystems. The unsustainable and poorly managed hunting has reduced some of the Greenland species. However, the government has taken some measures to obtain sustainable hunting. For instance, expert and scientific biological recommendations have been followed on the game species, increasing some populations. The regulations have also been implemented to control the harvest of several marine mammal species. However, there is still a need to take more actions to evaluate and control the harvest of nonregulated species (https://www.cbd.int/doc/world/dk/dk-nr-04-oth-en.pdf).

As the activities to extract natural resources within the Greenland area in the Arctic increase, it has become essential to regulate the navigation of the shipping responsible for carrying the natural resources from the Arctic to the rest of the world. The Kingdom of Denmark, in collaboration with other Arctic nations like the United States, has recommended the implementation of mandatory global and environmental safety laws for navigation in the Arctic. However, implementing these rules has not been successful due to conflict among the Arctic nations over the jurisdiction of the new Arctic shipping routes. Another problem associated with resource exploration and their export is that although Greenland is a part of the Kingdom of Denmark, it has self-governance, making it difficult to decide whether the Greenland government or the Denmark government controls the export of radioactive substances like uranium. Moreover, Greenland should consider that it should not violate the conventions and international treaties the Danish government has already signed while exporting uranium (https://www.lexology.com/commentary/energy-natural-resources/greenland/plesner/staking-a-claim-challenges-for-resource-extraction-in-the-arctic).

1.3 Consequences in the United States

There are potential oil spills that could occur in the Arctic region of the United States. So far, the highest amount of oil spilled in terms of pipeline leak was in the North Slope onto the tundra, specifically in the Prudhoe Bay field. This incident resulted in the release of approximately 267,000 gal of oil. The incident was because of corrosion-related factors and the negligence of the oil company BP, which was fined to the tune of USD 25 million. There is also the Exxon Valdez oil spill that prompted the development of oil spill management regulations and has been updated from time to time (Allison & Mandler, 2018).

There is also the impact of noise pollution and impact on the flora and fauna in this region. For example, the Beluga whale can get their habitat disturbed by noise and the release of oil in the water. Some of the other fauna that are impacted include bowhead and beluga whales, seals, walrus, polar bears, fish, birds, and caribou (USDC, 2011). Mining of precious minerals like gold, lead, and zinc is already taking place in Alaska, and the melting of ice would make it even more efficient. This is because transportation of equipment and personnel, as well as raw materials from the extraction, could be improved tremendously through Arctic shipping. On the other hand, however, the ice roads might begin to thaw and that may require a lot of money to maintain (UGCRTK, 2016). In terms of the impact of Arctic resource exploration and shipping, the main issues are oil spills polluting the flora and fauna and the impact on the Indigenous communities. The Indigenous people in Alaska already enjoy fewer health benefits than nonnative Alaskans and other US citizens. Overall, the Indigenous people in Alaska have a lower life expectancy ratio, a high risk of heart diseases, malnutrition, and a potentially high risk of infectious sickness. Activities related to resource extraction can worsen the health and quality of life of Indigenous people in Alaska (Moeser, 2023).

Alaska has an abundant resource of coal, which is not considered a clean fuel. About one-eighth of the global total coal reserves are discovered in Alaska. The massive amount of coal resources has attracted the attention of international mining companies, and several projects related to coal and mining, such as new strip mines and transport infrastructure development, have been started. The extraction and export of coal abroad will result in environmental pollution in Alaska and export destinations (https://alaskaconservation.org/protecting-alaska/priorities/protecting-lands-waters/energy-climate/). On the other hand, the exploration of natural resources like crude oil and natural gas in Alaska will have a minimum contribution to the economic growth of the US economy due to the small percentage of GDP consisting of the oil and gas industry compared to the rest of the sector. However, at the microlevel, these resources might positively impact the energy sector and significantly expand the production of products related to oil and petroleum (Nong et al., 2018).

1.4 Consequences in Finland

Finland is key to the European Union's (EU) quest to get the best of the Arctic. The town Oulu is very strategic in making this happen. This is because the main route would be through Russia and the West and Russia do not have the best relationship at the moment. There are plans underway to link Oulu to Northern Norway by rail. Northern Finland like many Arctic regions is not very developed, and so this presents a great opportunity for the area to benefit from the vast natural resource development in the area through mining (Confederation of Finnish Industries (CFI), 2022). Finland is an attractive place for world mining companies because of its rich resources and the Finnish government's support, such as reducing the tax rate, developing infrastructure, and providing financial support to the mining companies. However, these mining activities can negatively impact the lives of the Indigenous people. The Sami Parliament in Finland adopted no mining policy, and mining activities are restricted mainly to mechanical gold mining in Finnish Sapmi. Still, ongoing mining-related activities are the sources of threat, especially in Sami lands in the Kasivarsi region. Following are some other examples that resulted in conflicts with the Indigenous community: environmental damage due to the Talvivaara mine disaster in Eastern Finland; the discovery of the Mawson Resource gold in the Rompas-Rajapalo Natura region raises concern due to the efforts made against the gold mining in Kuusamo; and concerns raised by Finnish reindeer herders due to the execution of Yara phosphate mine project in Savukoski (Lassila, 2021).

There is also a huge opportunity for LNG transport. There has been interest in terms of investment from Japan, China, and other Asian countries (Hulac, 2015). The melting ice also presents Finland with more businesses since they are the leading icebreaker manufacturers in the world. In 2022, the Confederation of Finnish Industries issued a statement on the implications of the tariffs imposed by the EU on maritime transport. They among other things identified how this new tariff was going to make shipping activities more expensive in countries like Finland where they need icebreakers to navigate the ice-covered waters. The EU was charging more because of carbon dioxide emissions, which implies that countries like Finland that have a long navigating route and coast in the Arctic will be paying double for the icebreaker usage. They, however, suggested the following to help offset the cost: (i) Any extra cost incurred because of using ice-strengthened ships should be reimbursed by the EU, through appropriate instruments and institutions. (ii) The need to invest heavily in marine and low-energy carbon emissions should be intensified. (iii) Revenues from maritime emission trading should be reinvested in research and innovation in the maritime industry (CFI, 2022).

In a nutshell, the impact of Arctic shipping and resource exploration in the Finnish Arctic is more economical as opposed to social and to a lesser extent environmental. However, the environmental implications, especially carbon emissions, continue to be a major problem for most Arctic countries.

1.5 Consequences in Iceland

In the case of Iceland, it should be noted that it is located on the peripheries of the Arctic Circle, so the land mass per se may not be part of the Arctic (Protection of the Arctic Marine Environment (PAME), 2021). However, for the purpose of this chapter and subsequent discussions, we treat Iceland as an Arctic nation.

Even though some studies have suggested that Iceland may not be affected much by the opening of the Arctic shipping routes, others have disagreed, pointing to the huge number of resources in the Arctic basin, including oil and gas that could affect the local people and the Arctic marine environments (Keeling, 2020). Exploration licenses were issued for three oil and gas companies to prospect for hydrocarbon; however, almost two-thirds of them decided against going ahead with it. The country has also not designated sensitive areas for protection in case of oil spills (WWF, 2019). Since its establishment in 1996, Iceland has been a member of the Arctic Council and became chairperson in 2019 for 4 years. During its chairpersonship, Iceland prioritized sustainable development in the Arctic, including protecting the marine environment and Indigenous people and green energy outcomes. The government of Iceland has also developed strategies to protect the marine environment from plastic waste. None of the Indigenous tribes live in Iceland, but it still advocates for Indigenous people's rights, as reflected in its Arctic policy (https://www.thearcticinstitute.org/country-backgrounders/iceland/).

1.6 Consequences in Canada

To extract natural resources in the Canadian Arctic, offshore and onshore drilling projects are essential. However, the high exploration activities are due to offshore projects in the Beaufort Sea. From an economic point of view, resource extraction activities bring colossal revenue. Still, they have negative consequences such as environmental pollution, the probability of oil spills, and the well-being of Indigenous people (https://environment.geog.ubc.ca/energy-exploitation-in-canadas-arctic-solutions-to-a-wicked-problem-2/). Towns like Rankin Inlet have had their fair share of the impact of mining, especially nickel on the environment and health of the inhabitants. In fact, a recent survey reveals that residents want to know the potential implications of mining and shipping activities in the area before they are undertaken. The government has taken a bold step to try and reclaim land that has been polluted by nickel mining activities over the years

(Cater & Keeling, 2013). The Meliadine project, which is operated by Agnico Eagle for gold mining, is only the second by the company after the Meadowbank mine. The mine is approximately 25 km from Rankin Inlet and would provide a great economic opportunity to provide jobs for the inhabitants.

In 2022, the Canadian Coast Guard in Rankin Inlet received a boost of improved resources and investment to the tune of $24 million to convert their local station into an Arctic Marine Response station and another $84 million to map the ocean floor, all geared toward the safety of Arctic shipping and operations (CBC, 2022). Just like in Alaska the improvement of shipping in the Arctic would help supply food, personnel, and equipment to the region for mining purposes. Despite the positives, it is extremely important to assess the risks and design and develop measures to protect the inhabitant's health, safety, and culture. The social, economic, and cultural attributes significantly influence the health of people. Several social and economic variables affect the health of the communities, such as employment opportunities, business activities, work environment, access to quality food, housing, healthcare services, education, and good transport services. These factors affect the tendency of sickness and health care and consequently result in disparities in health conditions among the different segments of society. In British Columbia, the continuous growth and demand for resource extraction have contributed to inequalities in health in the whole province. Moreover, there is strong proof that resource exploration negatively affects social, economic, and cultural conditions, which harms the health and quality of life of people in northern Canada (Aalhus, 2018). The development projects in the Arctic pose a significant threat to the resources supporting Indigenous people's culture. To mitigate the risk to the culture, it is recommended that the government and policymakers should involve the Inuit in consultations and decisions, improve communication channels with native people, and listen to their opinions and concerns (https://environment.geog.ubc.ca/energy-exploitation-in-canadas-arctic-solutions-to-a-wicked-problem-2/).

1.7 Consequences in Norway

Norway enjoys the benefits of its geographical location within the Arctic, as the north of the Arctic covers 80% of Norway's seas. North Norway has a significantly higher economic productivity than the rest of the country. However, the Arctic area is also rich with several species and ecosystems that need protection. For instance, 15 million seabirds in the Barents Sea's surroundings and many polar bears and walruses are under threat due to vessel activities and resource extraction activities (https://www.openaccessgovernment.org/norways-arctic-strategy-innovation-sustainability/30391/). In Svalbard, Norway, there are plans in place to start drilling for oil in Arctic waters. However, critics say there are serious implications for the environment and its relationship with Russia, which is a signatory to the Svalbard pact (Sutterud & Ulven, 2020). Furthermore, earlier this year there was a proposal to undertake deep sea mining of rare earth minerals in the Arctic Ocean with the goal of satisfying the high demand for such in the emerging new green technologies. The Norwegian governments proposed the exploration of manganese, cobalt, lithium, gold, copper, zinc, and silver. A study of vessel activity in the Svalbard area shows that, over the years, the number and activities of fishing vessels have increased while passenger vessels remain constant (Stocker et al., 2020). This trend shows how the area is increasingly benefiting from the impact of climate change and the ease of shipping activities. Like Rankin Inlet, it is extremely important to have a plan in place to deal with potential oil spills should drilling go ahead as well as an economic plan to enable the inhabitants of the area to benefit fully from the resources being mined. As a part of Norway's Arctic policy, the Norwegian government has initiated and supported the project "Resource Extraction and Indigenous Peoples – Evaluating Ethical Guidelines." The project's objective is twofold: firstly, to formulate a framework for resource exploration. Secondly, to integrate Indigenous people to know their opinion about the performance of the business companies in the Arctic by suggesting a performance criteria system in which Indigenous people rank the business companies in terms of ethical fulfillment and building a collaboration with the native people. This strategy and other measures can give Indigenous people their rights and access to economic benefits (https://www.openaccessgovernment.org/norways-arctic-strategy-innovation-sustainability/30391/).

2 Discussions

From the review of the various countries, it is very common to see that most of the countries are worried about the prospect of drilling for oil in their region. This fear stems from the fact that when drilling for oil occurs there is also a potential for pollution of the Arctic marine environment. In addition, shipping of oil through the Arctic could also result in an accident and for that matter an oil spill. In some cases, these spills may affect the food sources of the inhabitants as well as their lifestyles and livelihoods like fishing and mining. In this case, mining communities are worried the waste from the mining activities may not be treated well and may eventually cause some diseases like cancer and kidney disease. Despite these challenges, most central governments continue to see the exploration of the Arctic resources as a way to improve their economy and develop new businesses. Promoting the region's political stability and collaboration among the Arctic countries are essential to gain economic benefits from these resources, which can minimize the fluctuation in energy prices. To protect biodiversity, Arctic nations should take measures to promote sustainable and well-managed hunting policies in the region. The Arctic countries should ensure that shipping and resource exploration activities do not harm Indigenous people's health, culture, and well-being and should provide a fair share of economic benefits generated from these resources to native residents, too.

3 Conclusions

In this chapter, the potential consequences of natural resource exploration in the Arctic as well as Arctic shipping were discussed. Oil spills, carbon emissions, impact on culture, and effect on the livelihood of the inhabitants were some of the identified impacts. The impacts in Norway, Canada, the United States, Iceland, Finland, Russia, and Greenland (Denmark) were discussed. Each country had its unique challenges. While environmental concerns were at the top of the worries for most countries, the development of oil and gas fields remains a major priority for Russia as it has also received a lot of support in terms of investments from China and other European countries through the Yamal LNG project. The Indigenous people of the Arctic region in the various countries are different, but they all face the same challenges, which is the threat to their way of life. Apart from oil and gas, other precious minerals' mining activity remains a huge prospect for this region. There is a divide between pursuing economic development in these areas and the environmental implications of engaging in such activities. Despite these divides, what to do about the huge resource potential of the area remains the prerogative of the individual countries.

References

Aalhus, M. (2018). *The social determinants of health impacts of resource extraction and development in rural and northern communities: A summary of impacts and promising practices for assessment and monitoring*. Northern Health and the Provincial Health Services Authority.

Afenyo, M., Khan, F., & Veitch, B. (2016). A state-of-the-art review of fate and transport of oil spills in open and ice-covered water. *Ocean Engineering.*, *119*, 233–248.

Afenyo, M., Ng, A. K. Y., & Jiang, C. (2019). Climate change and Arctic shipping: A method for assessing the impacts of oil spills in the Arctic. *Transportation Research Part D*, *77*, 476–490.

Afenyo, M., Ng, A. K. Y., & Jiang, C. (2021). A multi-period model for assessing the socio-economic impact of oil spills in the Arctic. *Risk Analysis: An international Journal*, *42*(3), 614–633.

Afenyo, M., Ng, A. K. Y., Jiang, C., & Lee, P. (2022). A Bayesian-loss function (BLF) hybrid model for assessing the marine liability regime for ship-source oil spills in the Arctic. *Risk Analysis: An International Journal*, *43*(2). https://doi.org/10.1111/risa.14025.

Allison, E., & Mandler, B. (2018). *Oil and Gas in the U.S. Arctic. Managing resources in an oil- and gas-rich but harsh and fragile environment*. American Geosciences Institute. https://www.americangeosciences.org/geoscience-currents/oil-and-gas-us-arctic. (Accessed 19 July 2023).

Browse, J., Carslaw, K. S., Arnold, S. R., Pringle, K., & Boucher, O. (2012). The scavenging processes controlling the seasonal cycle in Arctic sulphate and black carbon aerosol. *Atmospheric Chemistry and Physics.*, *12*, 6775–6798. https://doi.org/10.5194/acp-12-6775-2012.

Browse, J., Carslaw, K. S., Schmidt, A., & Corbett, J. J. (2013). Impact of future Arctic shipping on high-latitude black carbon deposition. *Geophysical Research Letters*, *40*, 16. https://doi.org/10.1002/grl.50876.

Cater, T., & Keeling, A. (2013). "That's where our future came from": Mining, landscape, and memory in Rankin Inlet, Nunavut. *Études/Inuit/Studies*, *37*(2), 59–82. https://doi.org/10.7202/1025710ar.

CBC. (2022). Rankin Inlet Coast Guard expands into Arctic marine response station. *CBC North*. https://www.cbc.ca/news/canada/north/rankin-inlet-coast-guard-expansion-1.6548289. (Accessed 2 August 2023).

Confederation of Finnish Industries (CFI). (2022). *Finnish business and trade unions: The competitive disadvantage of arctic shipping must be compensated*. https://ek.fi/en/current/bulletins/finnish-business-and-trade-unions-the-competitive-disadvantage-of-arctic-shipping-must-be-compensated/. (Accessed 31 July 2023).

Goldstein, M. A., Lynch, A. H., & Norchi, C. H. (2022). Conflict's impact raises costs for Arctic shipping and the climate. *Nature*. https://www.nature.com/articles/d41586-022-01538-w. (Accessed 3 August 2023).

Hanaček, K., Kröger, M., Scheidel, A., Rojas, F., & Martinez-Alier, J. (2022). On thin ice – The Arctic commodity extraction frontier and environmental conflicts. *Ecological Economics*, *191*, 107247. https://doi.org/10.1016/j.ecolecon.2021.107247.

Hulac, B. (2015). *Future Arctic: More mining, more shipping and more tourists*. Scientific American. https://www.scientificamerican.com/article/future-arctic-more-mining-more-shipping-and-more-tourists/#. (Accessed 3 August 2023).

Keeling, D. J. (2020). Transport geography in Iceland. *Journal of Transport Geography*, *89*, 102875. https://doi.org/10.1016/j.jtrangeo.2020.102875.

Korkmaz, Ö. (2022). Do oil, coal, and natural gas consumption and rents impact economic growth? An empirical analysis of the Russian Federation. *Resources Policy*, 77, 102739.

Lassila, M. (2021). The Arctic mineral resource rush and the ontological struggle for the Viiankiaapa peatland in Sodankylä, Finland. *Globalizations*, *18*(4), 635–649.

Li, M., Zhang, K., Alamri, A. M., Ageli, M. M., & Khan, N. (2023). Resource curse hypothesis and sustainable development: Evaluating the role of renewable energy and R&D. *Resources Policy, 81*, 103283.

Lifshits, S., Glyaznetsova, Y., Erofeevskaya, L., Chalaya, O., & Zueva, I. (2021). Effect of oil pollution on the ecological condition of soils and bottom sediments of the arctic region (Yakutia). *Environmental Pollution, 288*, 117680.

Litvinenko, V. S., Tsvetkov, P. S., & Molodtsov, K. V. (2020). The social and market mechanism of sustainable development of public companies in the mineral resource sector. *Euroasian Mining, 2020*, 36–41.

Moeser, E. (2023). *Decolonization and Arctic Engagement: A Critical Analysis of Resource Development in the US Arctic.* The Arctic Institute Arctic Collaboration Series. Available at https://www.thearcticinstitute.org/decolonization-arctic-engagement-critical-analysis-resource-development-us-arctic/.

Nong, D., Countryman, A. M., & Warziniack, T. (2018). Potential impacts of expanded Arctic Alaska energy resource extraction on US energy sectors. *Energy Policy, 119*, 574–584.

Protection of the Arctic Marine Environment (PAME). (2021). *Overview of low impact shipping corridors & other shipping management schemes.* Arctic Council. https://pame.is/document-library/pame-reports-new/pame-ministerial-deliverables/2021-12th-arctic-council-ministerial-meeting-reykjavik-iceland/788-overview-of-low-impact-shipping-corridors-other-shipping-management-schemes/file. (Accessed 31 July 2023).

Romasheva, N., & Dmitrieva, D. (2021). Energy resources exploitation in the Russian Arctic: Challenges and prospects for the sustainable development of the ecosystem. *Energies, 14*(24), 8300.

Stocker, A. N., Renner, A. H. H., & Knol-Kauffman, M. (2020). Sea ice variability and maritime activity around Svalbard in the period 2012–2019. *Scientific Reports, 10*, 17043. https://doi.org/10.1038/s41598-020-74064-2.

Sutterud, T., & Ulven, E. (2020). *Norway plans to drill for oil in untouched Arctic areas.*

Tulaeva, S. A., Tysiachniouk, M. S., Henry, L. A., & Horowitz, L. S. (2019). Globalizing extraction and indigenous rights in the Russian Arctic: The enduring role of the state in natural resource governance. *Resources, 8*(4), 179.

US GOVERNMENT Climate Resilience Tool Kit (UGCRTK). (2016). *Arctic development and transport.* https://toolkit.climate.gov/regions/alaska-and-arctic/arctic-development-and%20transport#:~:text=As%20sea%20ice%20declines%20in%20the%20Arctic%2C%20activities,traffic%20increases%20the%20risk%20of%20introducing%20invasive%20species. (Accessed 19 July 2023).

USDC. (2011). *Effects of oil and gas activities in the Arctic Ocean. Draft environmental impact statement.* https://media.fisheries.noaa.gov/dam-migration/arctic_seismic_deis_volume1_opr1.pdf. (Accessed 3 August 2023).

WWF. (2019). *WWF Arctic council score card. Iceland.* https://www.arcticwwf.org/our-priorities/governance/arctic-council-scorecard-2019/country/iceland/. (Accessed 3 August 2023).

Yildirim, S., Gedikli, A., Erdoğan, S., & Yıldırım, D.Ç. (2020). Natural resources rents-financial development nexus: Evidence from sixteen developing countries. *Resources Policy, 68*, 101705.

CHAPTER FIVE

Operational risks of Arctic shipping

1 Introduction

Operations in the Arctic involve activities such as oil and gas exploration and production, shipping, and mining, as well as fishing. Each of these activities has heightened risk profiles when they are being carried out in ice-covered waters (Afenyo & Caesar, 2023). The reasons are that the Arctic is a very harsh and remote environment and response efforts are normally slow and difficult as well. Increased shipping activities in the Arctic significantly depend on well-organized collaboration between companies exploring the resources and their shipping partners. To compensate for the grave operational risks in the Arctic and make their operation more protective and resilient, the stakeholders must alter and follow the appropriate supply chain management approaches (Gunnarsson & Lasserre, 2023).

The Society for Risk Analysis defines risk as a function of probability or frequency of the occurrence of an event, the scenario, consequences, and the strength of probability estimation. This definition means that even for low-probability and high-consequence events, the risk could be high. Most activities in the Arctic are classified as such and could even change with increased activity in this region. Currently, the probability of events occurring in the Arctic can be categorized mainly as low frequency; this is because accidents related to oil spills, etc., occur occasionally. As stated earlier, the frequency may increase if the right steps are not taken. Coupled with the increased traffic in this region, the frequency may move from low to moderate. For this reason, the International Maritime Organization (IMO), through the Polar Code, has proposed guidelines for the operations of ships in the Arctic region (Afenyo & Caesar, 2023). The ratification of this code, however, depends on individual countries.

In terms of the consequences, the numbers are high. Most of the major occurrences offshore for shipping, as well as oil and gas exploration, have

Safe and Sustainable Arctic Shipping Management and Development
https://doi.org/10.1016/B978-0-443-18849-7.00003-8
Copyright © 2024 Elsevier Inc.
All rights are reserved, including those for text and data mining, AI training, and similar technologies.

posed substantial financial implications. Illustrative examples include the Exxon Valdez oil spill, the Deepwater Horizon oil spill, the Mauritius oil spill, and quite recently the Siberian oil spill in Russia (Afenyo et al., 2017; Afenyo & Caesar, 2023). As alluded to in the previous chapter, Russia together with other countries like China and France are operating one of the biggest LNG plants in the world in the form of the Yamal plant. The plant was so successful that the second phase of the project is already underway. The transport of commodities like LNG only increases the frequency and potential consequences during operations. However, having in mind the famous Kolskaya oil rig sink near Sakhalin, of which about 50 people were originally feared dead, it created a lot of questions about the level of the response and how fast things could be from the Russians. If the Russians were not very timely in response, then it leaves a lot of questions about the other Arctic countries (Emmerson & Lahn, 2012). On the other hand, however, the project is also catering to the economic needs of the countries involved and the local people. Russia has invested heavily in Arctic infrastructure, making the consequences a little less. This means that for the same oil spill, for example, the response will be faster in Russia compared to other Arctic countries. Countries like Canada have rather taken the environmental protection approach, limiting Arctic activities to a small scale. The risk profile therefore for the same tanker going through the Canadian Arctic compared to the Russian Arctic would be slightly different.

Our ability to estimate the risk profile of activities in the Arctic would help in planning and investment, as well as designing emergency response measures. There are different methods for calculating the risk of activities both in open and ice-covered waters. This chapter reviews some of them. While different types of risks exist, this chapter focuses on operational risk. This type of risk is associated with the activities that take place in the Arctic region during operations. It would not include economic, environmental, or social risk. These are subjects of another chapter.

Some of the well-known methods for assessing maritime risk include the use of the risk matrix, using a combination of the fault tree (FT) and the event tree (Bowtie), Failure Mode Effect and Criticality Analysis (FMECA), Failure Mode and Effects Analysis (FEMEA), and Root Cause Analysis (RCA). The risk matrix uses a qualitative approach by employing a mechanism that categorizes into ranges less damaging, moderate, and critical. The matrix is not fixed and is built by either an individual or a group. There are two sides, that is, the frequency side and the consequence side. The difficulty is assigning a number to the event in the matrix. The matrix is assigned different colors normally with red being very critical, yellow moderate, and green meaning the event did not have devastating effects and so does not require any serious evaluations. Those events that fall in the critical zone really need a lot of work to mitigate their consequences and make sure they don't occur in the future. The proposed method in this chapter will demonstrate this later (see Fig. 2).

The Bowtie is another popular risk assessment tool combining an FT and an ET. The combination of the two looks more like a bow tie. FT is a tool that is used to investigate the cause of an accident. The causative factors are mainly at the bottom, and what they contribute to is at the top. The one at the top may further be a reason for another occurrence, and that is also on top of the previous. So, in summary, the immediate cause of an item is broken down into several root causes with each of the root causes also further broken down. However, each of these causes is connected and related by logic gates (Afenyo et al., 2017). The depiction of this is in Fig. 3 and will be discussed later. The ET on the other hand mainly models the consequences. So, while the FT gives us the frequency of occurrence, the ET helps us predict the impact of the events in monetary terms or any other entity like the amount of oil spilled, death of people, destruction of properties, etc.

Failure Mode Effect and Criticality Analysis is an inductive approach that links a series of causes of failure. It is an extension of FMEA. The latter is qualitative in nature, while the former is quantitative. Therefore, with FMECA you perform FMEA and go a step further to perform criticality analysis (Afenyo et al., 2017).

2 Operational risk factors in the Arctic

A study conducted by Panahi et al. (2023) led to researchers identifying the following as risk factors during Arctic shipping: (1) inability to communicate properly, (2) charts that are not up to date, (3) remote nature of the Arctic challenging search and rescue efforts, and (4) human error. Most of these factors also apply to oil and gas operations as well. The data for this study was gathered during a scientific expedition on board the Canadian icebreaker Amundsen. These factors were solicited from experienced captains on board the ship as well as other critical workers like the engineers. The main goal of the study is to understand how current practices manage to present a reasonable resilience front to dealing with issues of risk management during Arctic shipping, especially through the implementation of the Polar

Code. The authors employed the Fuzzy Analytical Hierarchy Process (FAHP) to develop a matrix. The matrix is made up of two key functional areas, importance and satisfaction. The outcome reveals that practices may be high on the importance quadrant but may not necessarily be the same on the satisfaction side of the matrix. This therefore shows the inadequacies of the current practices. A lot is changing now in terms of technology, and so the effect of some of these risk factors may be reduced to some extent. The charts can be improved tremendously by employing the latest state-of-the-art technology to build a system that would give very accurate coordinates and locate objects during shipping. However, this would require extensive mapping of the Arctic Sea floor. This effort is already underway but has been a little slow considering how traffic has increased in this region over the years. Some of these efforts started in 2004, and in 2007 a similar joint exercise involving the United States of America Department of State (USDS), the National Oceanic and Atmospheric Administration (NOAA), the University of New Hampshire, and the National Science Foundation (USDS, 2007) took place. In addition, in 2022, the Monterey Bay Aquarium Institute in partnership with Natural Resources Canada, Korean Polar Research Institute, and the University of New Hampshire conducted a mapping exercise in the Canadian Beaufort Sea (Paull et al., 2022).

The presence of icebergs, as identified by a study conducted by Afenyo et al. (2017), is another notable factor. These giant structures pose operational risks to both shipping and oil and gas activities in this region. It is very important that Arctic-going ships are ice-strengthened by adhering to the Polar Code of the International Maritime Organization (IMO). Another recent critical risk factor is cyber attacks by different actors. The actors range from government entities to private ones. A review study conducted by Afenyo and Caesar (2023) reveals how vulnerable the maritime cyber systems are to hacking. These activities have happened many times in open water. However, there is no well-documented attack in the Arctic. With the increased rate of Arctic shipping, that could be different. A single attack on any of the Arctic-going ships, oil rigs, and infrastructure would be devastating considering the remote nature of the Arctic and the lack of resources in this region.

In addition, a report released by Llyod (Emmerson & Lahn, 2012) identified the following as key risk factors: remoteness, electronic communication, climate change-related factors, weather, icing, and icebergs. Recently, a study conducted by Fu et al. (2023) used the Object-Oriented Bayesian Model (OOBN) to assess the navigational risk associated with navigation in the Arctic. They utilized data from Lloyd's intelligence and maritime accident investigation reports and concluded that the risk of a grounding accident remains the highest and most probable. Similarly, a study conducted by Khan et al. (2018), utilized the object-oriented Bayesian network to assess the operational risk of ships taking voyages along the Northern Sea Route (NSR). Further Domeh et al. (2023) proposed a conditional probability-based model for assessing operational risks of small fishing vessels in Arctic and Arctic-like environments. In doing so, we identified a gap in developing the conditional probability tables. They argued that since the experts that are used normally do not give the same weight to a combination of risk factors, instead, they suggest a weight-based system for uniformity. Khan et al. (2019) used an integrated Nagel-Schrekenberg (NaSch) model and a Bayesian network (BN)-based probabilistic approach to predict the density of convoys going through Arctic waters to reduce traffic and jams to the barest minimum. As pointed out by Cardenas et al. (2023), there are a lot of uncertainties related to geohazards in the Arctic. They therefore proposed specific considerations for analyzing geohazards in this region. In their study, they identified the following as posing serious hazards in the Arctic: (1) gas hydrates, (2) weak layers, (3) landslides, (4) ice gouging, (5) permafrost thawing, and (6) frost heave in addition to those identified earlier. They were very particular about the level of uncertainties involved in working in the Arctic due to these hazards. They therefore suggested that irrespective of what the situation is, it is extremely important that specific uncertainties are evaluated for each scenario and put forward reasonable assumptions. In addition, the amount of relevant data available can be a big factor in addressing this gap. Furthermore, the level of agreement of experts on the level of uncertainties and the clear understanding of the phenomenon addressed would also be an important factor. There are a few methods that attempt to address both epistemic and aleatory uncertainties. However, the BN that would be introduced later seeks to address both to some extent. It is important to note that we cannot completely get rid of uncertainties, but reducing them to a minimum is, ultimately, the goal of every manager.

3 Polar Code and its role in risk management

In this regard, the discussion of Arctic operational risk factors would not be complete without the examination of the role that the Polar Code plays. After its enactment and subsequent outdooring/implementation in 2017, critics of the code and the process that brought this to fruition argue

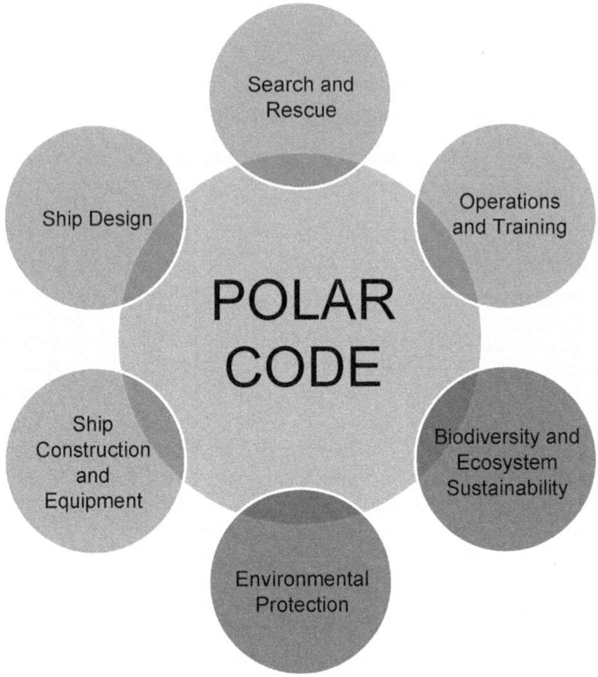

Fig. 1 The key themes of the Polar Code; the *blue* is operation related (SOLAS), and the *green* is more environmental protection related (MARPOL).

that the code provides a false sense of hope against dealing with the risk factors during shipping and other operations in the Arctic. The Polar Code is the working of over two decades of work, moving from voluntary rules to becoming one that has been adopted by the countries and companies around the world involved in Arctic operations. The Polar Code is the International Code for Ships Operating in Polar Waters enacted through the International Maritime Organization (IMO) (Fig. 1). Both the International Convention for the Safety of Life at Sea (SOLAS) and the International Convention for the Prevention of Pollution from Ships (MARPOL) were amended in 2014 and 2015, respectively, to accommodate the risks associated with navigation in the polar waters (International Maritime Organization (IMO), 2019). The adoption of the code means that ships hoping to operate in the Arctic and Antarctic waters would need to be certified under different categories (Table 1).

Despite the criticism, the Polar Code remains one of the only regulations available to manage operational risks in the Arctic. The main criticism stems

Table 1 Different categories under the Polar Code.

Category	Details
A	Can operate in medium first-year ice and in some cases old ice.
B	Can operate in thin first-year ice and some cases old ice
C	Can only operate in open waters

from the environmental side, for which activists argue that the provision within the Polar Code is not enough to address the environmental concerns related to Arctic shipping and operations. From ship design through construction to operations and training as well as search and rescue, the potential risks need to be considered every step of the way. Due to the increased risks in the Arctic, shipping companies have also improved how they hope to report routeing, incidents, and voyage planning (IMO, 2019).

4 The Bayesian Network as a tool for assessing the operational risk of arctic shipping

As alluded to earlier, many tools have been used to attempt to assess the operational risk of activities in the Arctic. However, to the best knowledge of the authors, none of them has a better advantage than the use of the BN and would form the basis of the demonstration of how to estimate the operational risk of Arctic activities. The BN is a probabilistic graphical-based network, consisting of nodes and arcs (Afenyo et al., 2017). The BN is rooted in the conditional probability theory, which among other things has the flexibility to become updated when new information becomes available. In addition, the theory helps in modeling dependency, which is very critical in the case of Arctic operations. This is because most risk factors for Arctic operations affect each other and are also dependent on each other. In the next section, a methodology is presented that describes how to develop the BN from raw data till it can be coded into an operational network. The theoretical basis for the BN is given in Eq. (1):

$$P\left(\frac{A}{B}\right) = \frac{P\left(\frac{B}{A}\right) \times P(A)}{P(B)} \tag{1}$$

where $P(A)$ is the prior probability, $P(B/A)$ is the likelihood, and $P(B)$ is the normalization factor. The BN is defined by $\langle Q, W \rangle$, where Q is the directed acyclic graph (DAG) with nodes X_1, X_1, ... X_n representing random variables. These nodes could be Boolean, for example, $\{\text{Yes}, \text{No}\}$,

Table 2 Sample applications of BN.

Work	Details	Area of application
Liu et al. (2022)	Examined the resilience of the Arctic shipping system against potential risks.	Safety of Arctic shipping
Harwood et al. (2021)	Integrated the BN theory in analyzing climatic variability in the Arctic	Teleconnection analysis
Xu et al. (2022)	Assess the potential for ship besetting during convoy operation during a voyage	Convoy operations
Li et al. (2022)	Addresses uncertainties in Arctic risk management for liquified natural gas (LNG) transportation.	Risk management
Aziz et al. (2019)	Assessing risks of operational systems on a marine vessel.	Risk management

values like {cold, moderately cold, extremely cold } and sometimes values like the age of girls {1......30} (Afenyo et al., 2017).

The use of BN in maritime operation applications is well documented, and a snapshot of some of the works is shown in Table 2.

5 Methodology

A proposed method is now presented on how to conduct the evaluation of operational risk during Arctic shipping and operations. To do this, the first step is to screen the web for Arctic shipping and related incidents. See Fig. 2. Also, there are databases from the governments of Arctic Council members that are available for these screening exercises. The next step is to further screen these databases for faults during operations. After, these faults are categorized into root cause, intermediate cause, and immediate cause. Normally, the root causes are the things that are key reasons for a particular incident. The intermediate cause on the other hand may not be a glaring reason but may be a symptom of a root cause. Finally, the immediate cause is what we all perceive as the cause that immediately leads to the occurrence of the incident.

This exercise is done with the help of the risk matrix shown in Fig. 3. The incidents considered in this exercise are those falling in the yellow and red regions. In this case, the red region is very critical, and the yellow region is critical but not as critical as the red region. The incidents that fall in the green region are not something we should worry about in the foreseeable future.

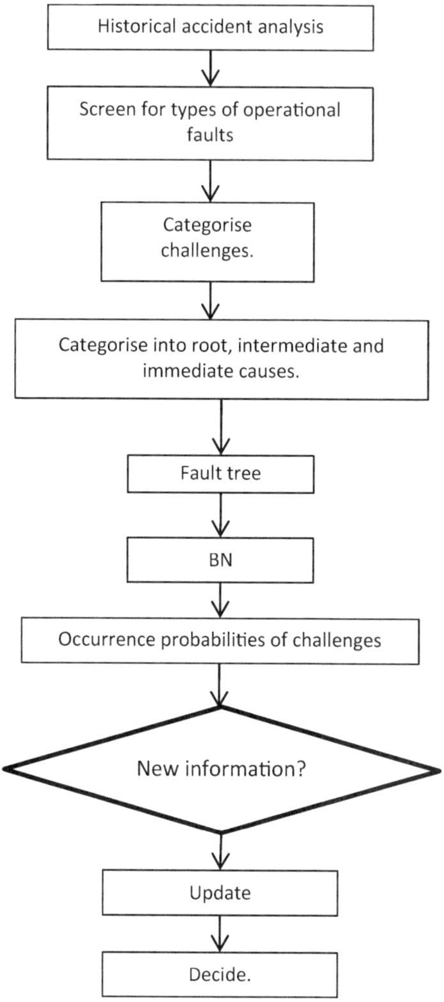

Fig. 2 The flowchart showing step by step the methods used. *Modified but adapted from Afenyo et al. (2017).*

S/F	1	2	3	4	5
1	1	2	3	4	5
2	2	4	6	8	10
3	3	6	9	12	15
4	4	8	12	16	20
5	5	10	15	20	25

Fig. 3 The risk matrix that could be used for screening the incidents during Arctic shipping operations.

These categorized incidents are used in building an FT. An example of this FT for ship collision with a fixed object like the iceberg is shown in Fig. 4. The FT uses the "AND" and "OR" gates in this case (Of course, there are more gates available but they are not discussed in this chapter.). The "OR" gate operates with the premise that only one factor needs to be a reason for a particular incident to happen. The "AND" gate uses the idea that both factors must be responsible for the incident or factor to occur. The weakness, however, of the FT is that it is not able to model dependencies and is not dynamic.

Therefore, the next step is taken to map the FT into a BN. This can be seen in Fig. 5 (See Afenyo et al. (2017) for further details.).

In the BN, human error is attributed to inexperience or distraction during navigation, slip and fall, sleeping and alcohol influence, or a combination of any of them. These may eventually result in the collision with a fixed object. On the other side, radar failure due to software or electrical malfunction may be the other immediate factor causing a collision with a fixed object like the iceberg. The Bayesian model gives us the probability of occurrence of that scenario using the factors gathered. To estimate the risk, it is important to multiply this by the consequence of the event. For example, if the event resulted in the damage of a certain value of materials, the product of the value and the probability gives the risk. The other factors that need to be considered when defining the risk are the scenario as well as the strength of probability.

A step further these days is not only to know the risk of a particular event but to determine how resilient the systems are to recover back to the original state they were.

6 Discussions

The discussions on the operational risk in the Arctic will continue to evolve, so will the methods and the tools to assess it. Hitherto, there is still inadequate data on some of these happenings. Hence, one of the best methods to use is our proposed BN that combines both qualitative and quantitative data. In addition, it is superior to other static methods available, which have already been discussed. Having a good risk management plan requires a corresponding risk assessment tool. Therefore, it is essential that we capitalize on the big data analytical tools that are available. Uncertainties are another critical issue that needs to be addressed, and to do this, the BN attempts to address both the epistemic and the aleatory uncertainties. Risk factors like bad communication instruments and outdated radar systems can

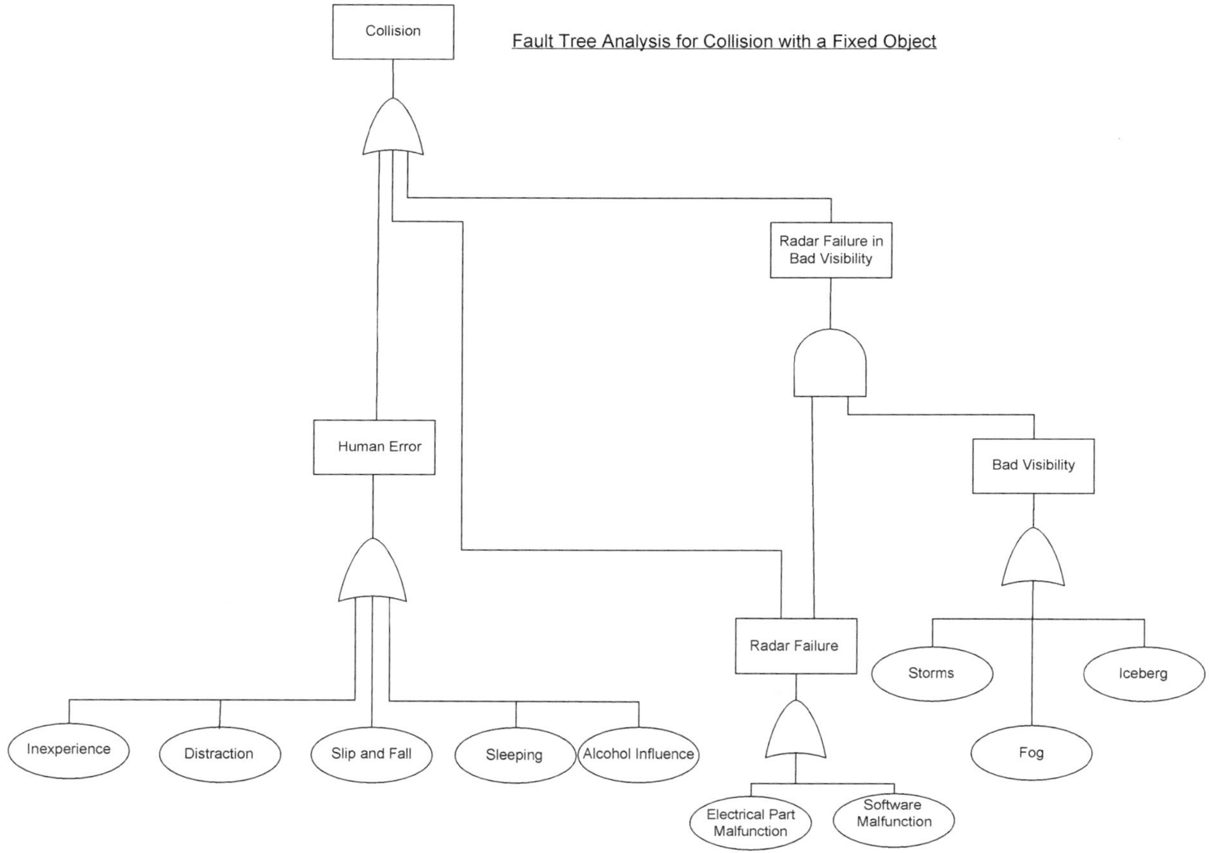

Fig. 4 Modified fault tree for ship collision with a fixed object (Afenyo et al., 2017).

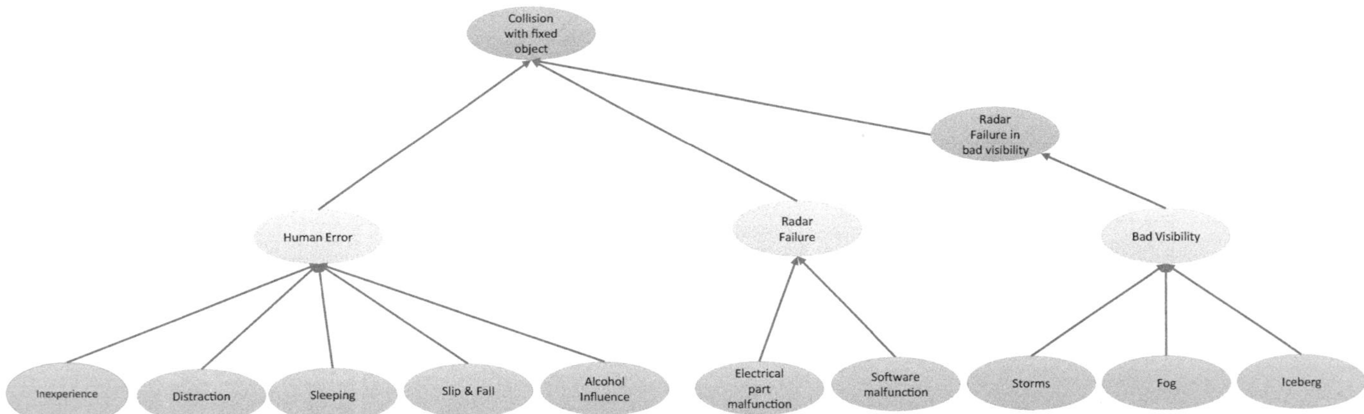

Fig. 5 The mapped Bayesian Network. *(After Afenyo, M., Khan, F. Veitch, B., & Yang, M. (2017). Arctic shipping accident scenario analysis using Bayesian network approach.* Ocean Engineering, 133, *224–230).*

be improved, while the size of icebergs may be reduced; however, risk factors such as human factors would remain a major problem as more sophistication is added to addressing the issue of operation of equipment. To make all these worries an issue across the Arctic world, the Polar Code needs to be updated to reflect the new realities of the Arctic, including addressing the issue of maritime cybersecurity. Any harking of the Arctic-going vessels and infrastructure may be difficult to address mainly because of the remoteness, harshness, and the lack of personnel available for such. However, having a proactive approach to this by incorporating these into the syllabus of the maritime training institutions around the world would be a great first step to making this a reality.

7 Conclusions

In this chapter, operational hazards during Arctic shipping and accompanying operations are presented. Also, the concept of risk was introduced and the role of the Polar Code for effective risk management in polar waters is presented. The type of risk that this chapter focuses on is operational risk. It identified a lot of challenges with regard to estimating the operational risks of this region, especially with regard to the lack of data and uncertainties. It therefore introduced the BN that addressed the dynamism of the Arctic and model dependency and had the capacity to get updated when new information becomes available. The Bayesian theory was used to propose a risk assessment method, which involved the use of a risk matrix, as well as a FT. However, operational risk assessment remains a challenge for terrains, such as the Arctic; modern statistical advancement, as well as improved ways of collecting data, can make things easier and better. Moving forward, we envisage improved accuracy in model predictions of the operational risks in the Arctic.

References

Afenyo, M., & Caesar, D. (2023). Maritime cybersecurity threats: Gaps and directions for future research. *Ocean and Coastal Management.*, *236*, 106493.

Afenyo, M., Khan, F., Veitch, B., & Yang, M. (2017). Arctic shipping accident scenario analysis using Bayesian network approach. *Ocean Engineering.*, *133*, 224–230.

Aziz, A., Ahmed, S., Khan, F., Stack, C., & Lind, A. (2019). Operational risk assessment model for marine vessels. *Reliability Engineering & System Safety*, *185*, 348–361.

Cardenas, I. C., Flage, R., & Aven, T. (2023). Marine geohazards exposed: Uncertainties involved. *Marine Georesources & Geotechnology*, *41*(6), 589–619.

Domeh, V., Obeng, F., Khan, F., Bose, N., & Sanli, E. (2023). An operational risk awareness tool for small fishing vessels operating in harsh environments. *Reliability Engineering & System Safety, 234*, 109139.

Emmerson, C., & Lahn, G. (2012). *Prepared for Lloyd's. Arctic opening: Opportunity and risk in the high north. Chatham House.* https://assets.lloyds.com/assets/pdf-risk-reports-arctic-risk-report-webview/1/pdf-risk-reports-Arctic-Risk-Report-webview.pdf. (Accessed 14 August 2023).

Fu, S., Zhang, Y., Zhang, M., Han, B., & Wu, Z. (2023). An object-oriented Bayesian network model for the quantitative risk assessment of navigational accidents in ice-covered Arctic waters. *Reliability Engineering & System Safety, 238*, 109459.

Gunnarsson, B., & Lasserre, F. (2023). Supply chain control and strategies to reduce operational risk in Russian extractive industries along the Northern Sea route. *Arctic Review on Law and Politics, 14*, 21–45.

Harwood, N., Hall, R., Di Capua, G., Russell, A., & Tucker, A. (2021). Using Bayesian networks to investigate the influence of subseasonal Arctic variability on Midlatitude North Atlantic circulation. *Journal of Climate, 34*(6), 2319–2335.

International Maritime Organization (IMO). (2019). *International Code for Ships Operating in Polar Waters (Polar Code).* https://www.imo.org/en/OurWork/Safety/Pages/polar-code.aspx. (Accessed 15 August 2023).

Khan, B., Khan, F., & Veitch, B. (2019). A cellular automation model for convoy traffic in Arctic waters. *Cold Regions Science and Technology, 164*, 102783.

Khan, B., Khan, F., Veitch, B., & Yang, M. (2018). An operational risk analysis tool to analyze marine transportation in Arctic waters. *Reliability Engineering & System Safety, 169*, 485–502.

Li, Z., Hu, S., Zhu, X., Gao, G., Yao, C., & Han., B. (2022). Using DBN and evidence-based reasoning to develop a risk performance model to interfere ship navigation process safety in Arctic waters. *Process Safety and Environmental Protection, 162*, 357–372.

Liu, Y., Ma, X., Qiao, W., & Han, B. (2022). A methodology to model the evolution of system resilience for Arctic shipping from the perspective of complexity. *Maritime Policy & Management.* https://doi.org/10.1080/03088839.2023.2187472.

Panahi, R., Afenyo, M., & Ng, A. K. Y. (2023). Developing a resilience index for safer and more resilient arctic shipping. *Maritime Policy & Management, 50*(7), 861–875.

Paull, C. K., Dallimore, S. R., Jin, Y. K., Caress, D. W., Lundsten, E., Gwiazda, R., Anderson, K., Clarke, J. H., & Melling, H. (2022). Rapid seafloor changes associated with the degradation of Arctic submarine permafrost. *Proceedings of the National Academy of Sciences of the United States of America., 119*(12), e2119105119.

USDS. (2007). *Mapping the Arctic Sea floor.* Bureau of Oceans and International Environmental and Scientific Affairs. https://2001-2009.state.gov/r/pa/ei/pix/b/sat/93321.htm. (Accessed 15 August 2023).

Xu, S., Kim, E., Haugen, S., & Zhang, M. (2022). A Bayesian network risk model for predicting ship besetting in ice during convoy operations along the Northern Sea route. *Reliability Engineering and System Safety, 223*, 108475.

SECTION C

Operational challenges of Arctic shipping

CHAPTER SIX

Arctic oil spills: Should we be worried?

1 Introduction

Oil spills are classified as low-frequency high-impact events and often are regarded as causing a lot of problems whenever they occur. Some of the problems range from health to environmental ones (Afenyo, Khan, Veitch, & Yang, 2016). Over the years, there have been major oil spills that have occurred both in the polar regions and the open seas. These spills have different causes and are mostly from ship operations and oil and gas exploration and development activities, with many of them routine spills during ship operations (Ng & Song, 2010). The illustrative examples of serious oil spills include the Exxon Valdez oil spill and the Deepwater Horizon oil spill (Afenyo et al., 2021; Afenyo, Khan, & Veitch, 2016). The former is a shipping accident while the latter is a well blowout. The causes of different types of oil spills vary, but the consequences remain largely the same. The causes of oil spills range from grounding, collision, and sinking of ships to malfunction of oil and gas instruments. Table 1 shows different oil spills that have occurred over the years and their impacts.

Another notable thing is that oil spills occur almost everywhere and anywhere there are maritime operations; the likelihood for an oil spill to occur exists. However, in both polar regions of the Arctic and the Antarctic, not much has been recorded in terms of oil spills generally compared to open waters. This is partly due to the length and intensity of maritime operations that have been going on compared to open waters. The notable one is the Norilsk oil spill in the Arctic and the Bahia Paraiso oil spill in the Antarctic, with the Norilsk oil spill believed to be caused by abnormal temperature change (Rajendran et al., 2021) (and, arguably, the effect of climate change). The effects of oil spills are devastating, and often communities and the ocean continue to experience the consequences many years after such spills. However, oil spills in general have one thing in common: They pollute the water

Safe and Sustainable Arctic Shipping Management and Development
https://doi.org/10.1016/B978-0-443-18849-7.00014-2
Copyright © 2024 Elsevier Inc.
All rights are reserved, including those for text and data mining, AI training, and similar technologies.

Table 1 Oil spills that have occurred over the years and their impacts.

Year	Name	Location	Impact
1979	Atlantic Empress oil spill	Off the coast of Trinidad and Tobago	90 million gallons of oil released
1989	Bahia Paraiso oil spill	Bahia Paraiso, Antarctica	60,000 L of marine diesel released into the sea
1991	Persian Gulf oil spill	Kuwait, Persian Gulf	380 to 520 million gallons
1994	Komi oil spill	Russia	79,000 t/58,5000 barrels released into a water body
2010	BP Deepwater Horizon oil spill	Gulf of Mexico	206 million gallons released into waterbody
2020	Norilsk oil spill	Norilsk, Russia Arctic	21,000 tons of oil entered waterways in the city. The groundwater was also polluted
2020	MV Wakashio oil spill	Mauritius	1000 metric tonnes of oil released into the environment

bodies and, in some cases, render people out of jobs. Also, they cause social disruption and economic hardship and affect the culture of the people, especially if it occurs in the polar regions. The economic, social, and environmental impacts of the spills will be further discussed in Chapters 4 and 9.

As the Arctic has become relevant these days because of more intense transportation and mineral exploration, the potential for oil spills is as high as ever. A major oil spill in the Arctic would be devastating and may take years to clean up, and probably we might never reach some of the oil as it could become engulfed in ice. In addition, tracking the spilled oil is a huge challenge thanks to the harsh and remote nature of the Arctic area. No matter the causes, of oil spills, the government and other organizations for the past decade have engaged in a lot of research to make sure that we are well prepared for potential oil spills as much as possible. In this regard, one such example is the Arctic Response Technology JIP (see https://www.arcticresponsetechnology.org/mediaroom/arctic-oil-spill-response-technology-jip-summary-report/). In this research, they examined the fate and transport of oil in ice-covered waters, environmental impacts of an oil spill in ice-covered waters, response technologies available for addressing Arctic oil spills, and the policies and regulations governing oil spills. The outcomes were promising, but the authors cautioned that the tools alone may not be enough and need research to improve their capabilities before we are ready for the major oil spill in the Arctic region.

The major processes that occur after an oil spill in the Arctic are like those in open water, except that there are other additional processes like encapsulation, which is specific to just ice-covered waters. In addition, how oil behaves in ice is different from how it is in open water. In the next section, we will examine some of the processes involved when an oil spill occurs and why we should be very worried when it happens in ice-covered waters. Also, despite the regulations available for shipping in polar regions in terms of oil spills, it is rather alarming that there are no such regulations for upstream petroleum activities. In fact, hitherto, the Deepwater Horizon oil spill incident and the Ekofisk Bravo incident remain some of the very worrying incidents. When there is an oil spill from upstream petroleum activities, the impacts are even more as the flow is continuous and only stops after it has been capped, and, in certain cases, it becomes a full-blown oil spill incident (Law, 2011; Afenyo, Khan, & Veitch, 2016).

2 Fate and transport processes of oil

When an oil spill occurs, it undergoes a lot of processes that can be categorized as weathering and transport processes. These processes transform the oil into forms that are either soluble in the water or make them disappear. Weathering is generally defined as a combination of both physical and chemical processes that makes the oil slick disintegrate into a form that becomes a bit heavier than the water it was spilled in (Afenyo, Khan, & Veitch, 2016). Some of these processes happen immediately when we have an oil spill, while others happen later. Some depend on each other for the process to occur, while others do not. There are weathering processes such as evaporation. This involves the lighter or more volatile components of the oil spilled disappearing into thin air when exposed to wind and temperature. This volatile substance turns into vapor and leaves the surface. When this happens, the weightier components may further transform and sink to the bottom of the water body where the spill occurred (Afenyo et al., 2019). Depending on the type of oil we are dealing with, the rate of evaporation could be faster than others. For example, products such as gasoline (and to some extent kerosene) contain rather light ends. Their rate of evaporation is very fast and therefore poses minimum risk to aquatic fauna and flora. On the other hand, heavier oil leaves thicker residue and does not evaporate quickly. One of the weathering processes is oxidation, which results in water-soluble compounds. This happens when oil encounters water and oxygen. The process of oxidation may result in a substance often referred to as tar balls. Tar balls are difficult to deal with and may remain in the aquatic environment for a very long time (Fingas, 2015).

Biodegradation is another important weathering process and involves the breakdown of the oil slick by special oil-eating bacteria. Not every microorganism can do this, and therefore very specialized ones are responsible for the breakdown. These microorganisms need nutrients, such as phosphorus and nitrogen, to make this work efficiently. Even though the process of biodegradation is more effective in open waters (especially warmer waters), in the Arctic region there are specific microorganisms that have similar abilities as well. Emulsification is another very important weathering process. In fact, the process of emulsification results in a substance called emulsion, which is basically a combination of droplets of oil in water. Like dispersion, waves are responsible for initiating this process. Emulsion generally presents a huge challenge to cleaning efforts. The emulsion is sometimes called a chocolate mousse. This product may remain in the environment for a very long time posing health and other environmental problems to the people and the fauna in the water body (Afenyo, Khan, & Veitch, 2016; Fingas, 2015). Finally, there is photooxidation that uses light to transform oil slicks.

Transport processes, such as spreading, involve the movement of the oil from one point to the other. Encapsulation of the oil by ice entails the oil getting engulfed in the ice and in some instances gets transported miles away and later released into the water body when the temperatures increase. In this regard, dispersion can be both natural and artificial. Naturally, waves may start the process by disintegrating an oil slick into droplets and subsequently distributing them in a vertical way in the water column. In addition, the outcome of this process of natural dispersion can result in a thin film that can be seen on the surface of the water. Artificially, dispersion can be initiated and facilitated using dispersants, which will be further discussed later in this chapter.

Oil released in ice-covered waters could make its way into ice, under ice, in between leads, and on snow. Each of these situations presents extra complexity to the way oil will interact with its icy environment. Understanding such, it is very difficult to build models to accurately predict the behavior of oil as well as where the oil will go. In this case, the Norwegian research group SINTEF is a pioneer in trying to deal with this complexity. Their initial models have seen massive improvements over the years. Following it, other countries have also gone ahead to build their own in-house models to manage oil spills in their jurisdiction. Fig. 1 shows the processes that occur after an oil spill in open and ice-covered waters. As you can infer from Fig. 1, if the oil spills in snow, first- or second-year ice, the behavior of the oil will be different. This behavior of oil in ice makes it difficult to predict and track oil spills in icy waters.

3 Response to oil spills

Like oil spills in the open waters, there is a suite of tools available for response anytime there is an oil spill in the Arctic. Some of these include the use of dispersants, mechanical recovery, in situ burning, and monitored and natural attenuation (Afenyo et al., 2023). The mechanical recovery suit is sometimes made up of booms and skimmers and in some instances barriers. As alluded to earlier, response to oil spills in the Arctic becomes extremely difficult, and so scientists and other government organizations are employing modern tools like the use of radar and satellites to improve the efficiency of identification of spills and subsequent cleaning and management of it (Tunnicliffe, 2019). Various Arctic member countries have different approaches to response to oil spills. For Canada, the approach is to first contain the oil and then collect it. This is normally done using mechanical tools such as booms and skimmers and sorbents (Transport Canada (TC), 2022). However, the question remains on how such methods can be effective in a place like the Arctic, and the answer is that authorities might not be able to even deploy this tool to such areas. For example, Canada is not particularly comfortable with the use of both dispersants and in situ burning as they cause more pollution to the environment. Environment Canada, Transport Canada, Fisheries and Oceans Canada, and Natural Resources Canada are all involved in dealing with oil spills to some extent in Canada. Hence, the Canadian government and many other agencies have been looking at Monitored Natural Attenuation (MNA) (Afenyo et al., 2023) through collaborative research projects. An illustrative example is the GENICE project (see: genice.ca), a collaborative project funded by Genome Canada and led by Canadian researchers in addressing the socioeconomic, chemical, and biological impacts of oil spills in Canada's Arctic region, as well as the development of environmentally friendly solutions in addressing it.

In Norway, the Norwegian Coastal Administration under the Ministry of Transport and Communication takes care of anything related to oil spill management in terms of preparation and response on the national scale. On the other hand, the Ministry of Climate and Environment is responsible for

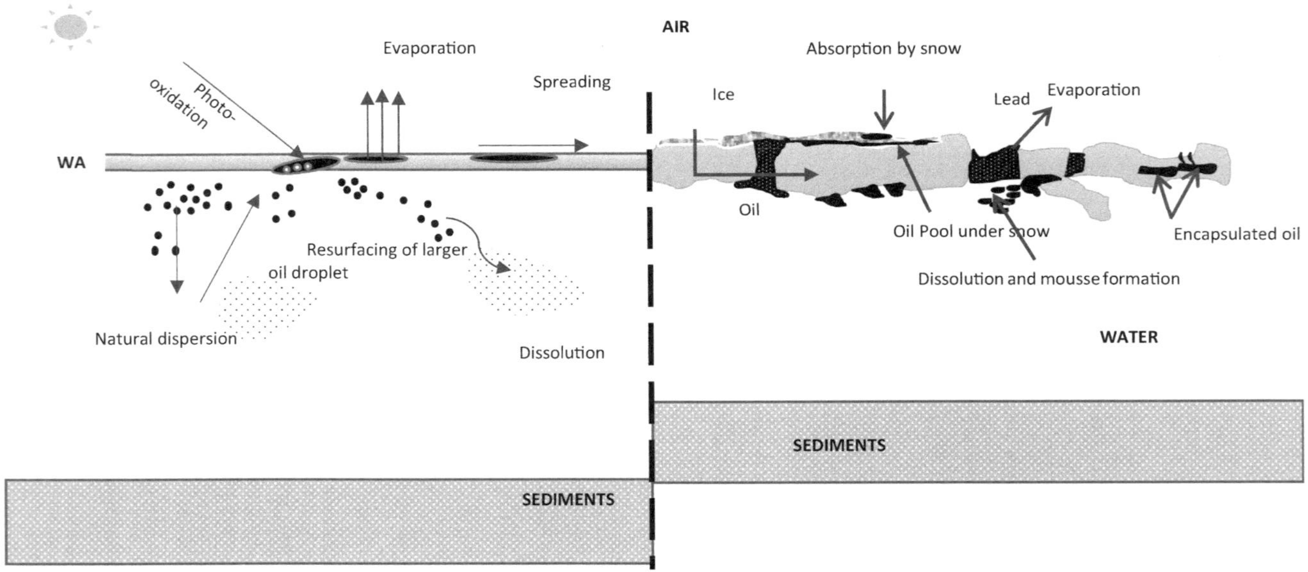

Fig. 1 Oil weathering in open and ice-covered waters. *(After Afenyo, M., Jiang, C., & Ng, A. K. Y. (2019): Climate change and Arctic shipping: A method for assessing the impacts of oil spills in the Arctic. Transportation Research Part D: Transport and Environment, 77, 476–490.)*

setting the standards for pollution from oil spill preparedness on the local scale mainly involving municipalities and private organizations. The Coastal Administration receives all reports on oil spills and subsequently proceeds to analyze and make recommendations on what to do (Norwegian Petroleum (NP), 2023). On the other hand, the approach the Finnish authorities have adopted toward oil spills in ice-covered waters is to wait until the ice melts and then deal with the oil through recovery techniques discussed earlier. If the oil spill is small in quantity, the Finnish authorities have the capacity to deal with such, while they only wait when the quantity is large. The Ministry of the Interior of the Finnish government is responsible for dealing with oil spills in Finland. The Finnish Boarder Guard and the Rescue Service Department are responsible specifically for dealing with spills on the national level and local levels, respectively. Like Canada, Finland does not allow the use of dispersants and is a signatory of the Helsinki Convention that allows the use of certain chemicals under very strict restrictions (ITOPF, 2023). Finally, in Russia, the frequency of oil spills is higher than in other oil-producing countries (Kondratenko, 2021). A study conducted by Belkina and Sarkova (2015), which compared the Norwegian oil spill response policy to that of the Russian one, concluded that Russia had a more prescriptive approach (contrary to the performance-based approach employed by, for example, Norway). However, both approaches require performing certain levels of risk assessments.

4 Policies regulating oil spills

As stated earlier, there are standards or rules for Arctic shipping but not for offshore oil and gas development. Having standards for oil and gas exploration and development in the Arctic would help in so many ways. First, it would provide guidelines for designing, transporting, and installation of equipment for safe operation in the Arctic region. In addition, it would help in the designing of contingency plans and response for oil spills from oil and gas operations in this region. Such a guide should consider the harsh weather, extremely low temperature, and the remoteness of this region.

A recent study conducted by the Pew Thrust regarding the US Arctic regions suggested that any facilities to be deployed and used in the Arctic should be designed in such a way that they could withstand extreme ice forces and sea states. Also, it suggested that the equipment for oil spill response should be designed with the conditions of Alaska and other Arctic regions in mind and must be in such a way that it could deal with and recover oil spills in ice-covered waters (The Pew Charitable Thrust (TPCT), 2013). This is important because, unlike the open waters, some of the oil could be under ice, in between the ice, or gets encapsulated by ice. Any systems in this region should always have a backup, given the remoteness of the region, and thus could not be visited as often as in the open waters and lands. In this regard, it might be important for researchers to begin investigating how we could effectively use artificial intelligence (AI) to deal with the risks and other happenings in the Arctic region. At the same time, as drones and other critical instruments and equipment could be operated remotely, we strongly recommend oil and gas companies, shipping companies, and other Arctic stakeholders consider investing heavily in such new technologies to make operations in this area less risky and more efficient.

Hitherto, the UN Convention on the Law of the Sea (UNCLOS) is one of the most comprehensive regulations to deal with operations in the maritime territories. It clearly states the duties of each state and what is expected of them. However, it relies heavily on the cooperation of individual states. In some cases, individual states have their own national regulations that could lead to clarity problems and create confusion about what to do when an oil spill occurs in international waters. On the other hand, there are other regulations that deal with oil spills but not Arctic specific, such as the International Convention on Oil Pollution Preparedness and the Preparedness and Response in the Arctic (MOSPA) (Shapovalova, 2019). These regulations will be further discussed in Chapter 3.

5 Discussions

Oil spills occurring in open waters are a serious concern since their effect can still be experienced many years after the incident. Even if the oil is recovered and cleaning is done, the impact continues to linger on for many years. Indeed, there are environmental, social, and economic implications. Should a similar incident of oil spill occur in ice-covered waters, it becomes even more complex. As explained earlier, oil in ice is extremely difficult to track and model, understanding that the spilled oil gets into the ice and follows different paths. In addition, processes that occur after an oil spill are equally (if not more) difficult to model and predict. This is the case because the processes may occur at the same time and some dependent on each other. Even more challenging is the fact that some do not have the algorithms that can accurately model them: Even if we manage to develop computer models to describe the behavior of oil in ice, most likely, the accuracy would be questionable with high uncertainties and low predictability power.

Another reason we need to be worried about oil spills in the Arctic region is the fact that the currently available response tools and methods are unable to deal with oil in ice effectively. Some of the methods that have been researched to be effective in ice-covered waters are, unfortunately, deemed not environmentally friendly enough, and, hence, certain Arctic countries like Canada and Finland have chosen not to deploy such methods. The main reason lies in the environmental implications of these methods. For example, the use of in situ burning further produces more carbon dioxide (CO_2) into the atmosphere, while the use of dispersants introduces chemicals into the water bodies. Many countries that are signatories to the Helsinki Accord prohibit the introduction of certain chemicals in some proportions into water bodies. Moreover, the current regulations available for dealing with oil spills in the Arctic are, to say the least, ambiguous. The question therefore is: What happens when we have an oil spill in international waters? How ill coordination happens, and who will be held responsible? In the Arctic where there is already dispute over sovereignty over certain parts of the region, it might be even more worrying and disturbing.

Furthermore, it is very disturbing that, unlike shipping, oil and gas exploration and development is not currently regulated properly in the Arctic region. There are no guidelines and standards for companies and individuals operating in this region to follow. A part of the reason may be that we still have very limited knowledge of the Arctic region. Until now, few oil spills have occurred in the Arctic, and thus not much is known about the impacts in the short and long terms. Recently, we witnessed an influx of new technologies coming on board in other sectors that could be used to address the Arctic remoteness and its harsh environment, where the use of unmanned aerial systems (UASs) serves as an illustrative example. Hence, over time, we might gradually address these challenges, but until then, there are still many valid reasons for us to worry about this topic.

6 Conclusions

In this chapter, the question of whether we should be worried about oil spills in the Arctic or not is addressed. We explore this from the processes involved when an oil spill occurs, available recovery methods of the oil in the icy waters, regulations, as well as the governmental support.

Our conclusion is that, until now, we still need to be worried as we are not adequately well equipped to deal with a large-scale oil spill in the Arctic region. There are some promising methods but are, unfortunately, not deemed environmentally friendly enough to be widely adopted by Arctic stakeholders and authorities. Despite such, there is research in progress to develop environmentally friendly dispersants. In the foreseeable future, it is very important that companies and governments invest in developing new technologies that can help operations in the Arctic to overcome its remoteness and harsh environment. The development of the drone is a perfect example of this. Also, there is an urgent need to establish an international protocol to deal with potential oil spills in international Arctic waters. Such an instrument would help clear some of the ambiguities in the current laws and regulations that we have. We certainly need to be much better prepared for oil spills occurring in ice-covered waters any season of the year. Finally, international collaborative networks between scholars, policymakers, and practitioners in addressing this issue could play pivotal roles in improving the efficiency of dealing with such a challenge. In this case, the establishment of the International Forum on Climate Change Adaptation Planning for Port, Transportation Infrastructure, and the Arctic (CCAPPTIA) (see www.ccapptia.com) serves as an excellent example.

References

Afenyo, M., Jiang, C., & Ng, A. K. Y. (2019). Climate change and Arctic shipping: A method for assessing the impacts of oil spills in the Arctic. *Transportation Research Part D: Transport and Environment*, *77*, 476–490.

Afenyo, M., Jiang, C., Ng, A. K. Y., & Lee, P. T. (2023). A Bayesian-loss function-based method in assessing loss caused by ship-source oil spills in the arctic area. *Risk Analysis: An Official Publication of the Society for Risk Analysis*, *43*(8), 1557–1571. https://doi.org/10.1111/risa.14025.

Afenyo, M., Khan, F., & Veitch, B. (2016). A state-of-the-art review of fate and transport of oil spills in open and ice-covered water. *Ocean Engineering*, *119*, 233–248.

Afenyo, M., Khan, F., Veitch, B., & Yang, M. (2016). Dynamic fugacity model for accidental oil release during Arctic shipping. *Marine Pollution Bulletin.*, *111*(1–2), 347–353.

Afenyo, M., Ng, A. K. Y., & Jiang, C. (2021). A multi-period model for assessing the socio-economic impact of oil spills in the Arctic. *Risk Analysis: An international Journal.*, *42*(3), 614–633.

Belkina, N., & Sarkova, O. (2015). Regulatory approaches to oil spill response in Norway and Russia. *Polar Geography*, *38*(1), 1–21. https://doi.org/10.1080/1088937X.2014.980451.

Fingas, M. (2015). *Handbook of oil spill science and technology*. John Wiley & Sons Inc. https://doi.org/10.1002/9781118989982.

ITOPF. (2023). *Country and territory profiles. Finland.* https://www.itopf.org/knowledge-resources/countries-territories-regions/finland/#:~:text=Response%20Policy&text=Finland%20has%20been%20investing%20in,the%20oil%20with%20traditional%20techniques. (Accessed 6 September 2023).

Kondratenko, T. (2021). *Russia: A country of oil spills.* DW Nature and Environment. https://www.dw.com/en/russia-oil-spills-far-north/a-56916148. (Accessed 6 September 2023).

Law, R. J. (2011). The Ekofisk bravo blowout, 1977. In M. Fingas (Ed.), *Oil spill science and technology* (pp. 1107–1108). Gulf Professional Publishing. https://doi.org/10.1016/B978-1-85617-943-0.10034-6.

Ng, A. K. Y., & Song, S. (2010). The environmental impacts of pollutants generated by routine shipping operations on ports. *Ocean and Coastal Management, 53*(5–6), 301–311.

Norwegian Petroleum (NP). (2023). *Acute Pollution and oil spill preparedness and response.* https://www.norskpetroleum.no/en/environment-and-technology/oil-spill-preparedness-and%20response/#:~:text=The%20Norwegian%20Environment%20Agency%20approves,and%20the%20causes%20are%20investigated. (Accessed 6 September 2023).

Rajendran, S., Sadooni, F. N., Al-Kuwari, H. A. S., Oleg, A., Govil, H., Nasir, S., & Vethamony, P. (2021). Monitoring oil spill in Norilsk, Russia using satellite data. *Scientific Reports, 11*, 3817. https://doi.org/10.1038/s41598-021-83260-7.

Shapovalova, D. (2019). *Can international law protect the Arctic from oil spills?.* https://www.thearcticinstitute.org/international-law-protect-arctic-oil-spills/. (Accessed 6 September 2023).

The Pew Charitable Thrust (TPCT). (2013). *Arctic standards: Recommendations on oil spill prevention, response, and safety.* Pew. https://www.pewtrusts.org/en/research-and-analysis/reports/2013/09/23/arctic-standards-recommendations-on-oil-spill-prevention-response-and-safety. (Accessed 6 September 2023).

Transport Canada (TC). (2022). *Intentions paper: Enhancing oil spill response – alternative response measures.* Government of Canada. https://tc.canada.ca/en/corporate-services/consultations/let-s-talk-responding-oil-spills-alternative-response-measures/intentions-paper-enhancing-oil-spill-response-alternative-response-measures. (Accessed 5 September 2023).

Tunnicliffe, A. (2019). *Eye in the sky-the role of drones in oil spill management.* Offshore Technology. https://www.offshore-technology.com/features/eye-in-the-sky-the-role-of-drones-in-oil-spill-management/?cf-view&cf-closed (Accessed 7 September 2023).

CHAPTER SEVEN

Arctic shipping insurance

1 Introduction

Insurance is an integral part of shipping, and recently, it has gained much more attention in the context of Arctic shipping (Afenyo et al., 2022). Insurance depends much on assessing the risk of potential accidents and accompanying consequences. For a place like the Arctic where the traffic is on the ascendency, it is necessary to establish a way to assess the insurance premium of Arctic-going ships. Indeed, the shipping industry is currently facing substantial challenges assessing the premium for Arctic-going ships. In addition to Arctic transportation, oil and gas activities is another that requires equal attention. The premier international organization for marine insurance, The International Union of Marine Insurance (IUMI), has identified the following as key challenges to assessing risks in the Arctic, namely: (i) the lack of infrastructures and (ii) little provision for rescue should an accident occur. In this regard, Theocharis et al. (2019) has conducted a study exploring the feasibility of using the Northern Sea Route (NSR) as an alternative to the Suez Canal for Asia-Europe shipping routes. They developed a model that considered: (i) the distance to be covered by the ship, (ii) the type of oil that the tanker is carrying and (iii) the size of the ship, (iv) ice-breaking fees, and (v) environmental policy implications. They argue that these factors are key to assess the risk of operating in Arctic waters and subsequently developing an insurance premium for it. These factors are not only applicable to NSR, but could also be relevant to other major Arctic shipping routes, such as the Northwest Passage (NWP).

The famous case of *Clipper Adventurer*, which was carrying approximately 100 passengers when it ran aground is a classic example of an accident from shipping in the Arctic. The incident resulted in a series of suits between the Canadian government and the ship owners. While the ship owners sued the Canadian government for approximately USD13.5 million, the Canadian government counter-sued for polluting the ocean. In the end, the Canadian

Safe and Sustainable Arctic Shipping Management and Development
https://doi.org/10.1016/B978-0-443-18849-7.00006-3
Copyright © 2024 Elsevier Inc.
All rights are reserved, including those for text and data mining, AI training, and similar technologies.

government was victorious and was awarded USD469,000 (Hinchey, 2017). This case laid bare the gap in terms of who should be held accountable for accidents in the Arctic. As rightly pointed out by Panahi et al. (2020), there are risks such as those related to extreme weather events, which need to be taken very seriously when dealing with such risks. In the Arctic, communication is very important to making sure that oil spill contingency plans (OSCP) are implemented. In a study conducted by Grabowski et al. (2016), the authors emphasized the challenges of dealing with risks in the Arctic and resource allocation is one of the main issues. This is because in places like the Arctic, the presence of ice remains a major problem. All these make ensuring ships going to the Arctic a major challenge.

A study conducted by Wang et al. (2018) evaluated how insurance premiums could have an influence on the ex-ante declared value and the ex-post compensation. They presented a model that seek to raise the efficiency of the transport processes. The essence of such a model is to help customers decide if it is worth buying an insurance or not. The operational environment of a particular distributor contributes to the strategies for Business Interruption (BI) and a transport contingency strategy. A study was done to link the two and examine the choices that vessel owners and passengers have when they decide a particular insurance scheme. It is thus critical that vessels sailing in Arctic waters have diverse choices in choosing from different insurance schemes (Zhen et al., 2016).

In Fig. 1, the disparity between the requested claim and what is actually paid is presented in Canada. Although we should be cautious about generalizing the Canadian case for all countries and industries, it does illustrate the potential disparity that could exist between the two attributes.

Since there are many uncertainties concerning claims from the affected and how much insurance companies should be charging for their services, it is important to have a fair and acceptable premium. In the Canadian Arctic, for example, the burden of loss due to oil spills is on the affected party and so often difficult to evaluate (DeCola, 2020). This is because it is extremely difficult to monetize the non-economic losses on cultural factors, such as the disintegration of local families.

Furthermore, the blockage of the Suez Canal by the *MV Ever Given* in March 2021 and the recent Red Sea crisis since October 2023 illustrate the importance of the Arctic as an alternative to global shipping. Hence, it was little surprise that recent research has highlighted the need to urgently develop the Arctic shipping routes (e.g., Dixon (2021)). As an illustrative example, the blockage of the Suez Canal by the *MV Ever Given* consequently

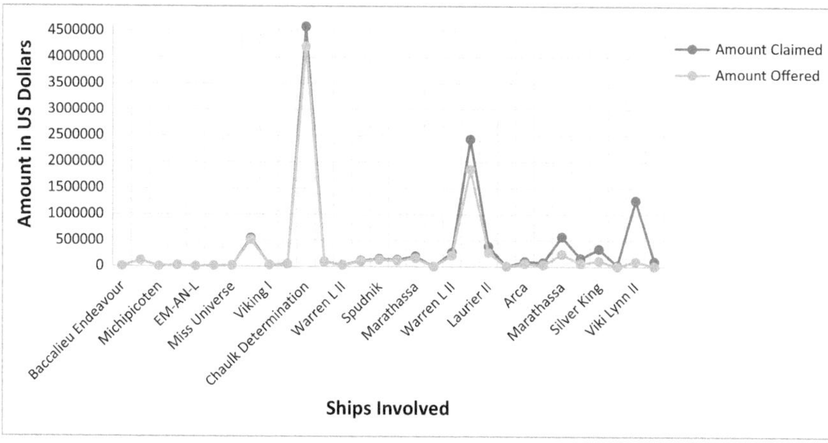

Fig. 1 Amounts claimed against among requested in selected ship-source oil spills in Canada. *(Source: Authors based on Anon. (2019). Prepared for WWF. WWF Canada's final written submission for the Nunavut Impact Review Board's reconsideration of the Mary River Project Certificate for Baffinland's Mary River "Phase 2" Proposal.)*

reduced global trade by an amount of approximately 12%. This follows on the back of a \$15 M loss of revenue by the Suez Canal Authority during this incident (Russon, 2021).

Currently, the practice is to add 40% of the basic premium used in open waters to the original amount charged. It means clients pay 40% extra for navigating in ice-covered waters. This figure, however, might vary from fleet to fleet, depending on experience, type of ship and weather, and other environmental conditions. This method is not only problematic to the clients but also to the insurance companies. It is not very accepted in the industry, and there is a need to have some objective way to develop a fair and realistic risk premium. The budding tourism industry in the Arctic may also face major challenges if a realistic risk premium is not developed. With the ice melting at an alarming rate, the tourism industry in the region is likely to improve tremendously.

To fill this gap, the chapter seeks to explore methods that could be used to develop such a premium.

2 Literature reviews

There is literature on the risk of Arctic shipping but very limited on the insurance of shipping in the Arctic. This section brings into perspective some of the related works on insurance of shipping. In a study conducted by

Johannsdottir and Cook (2019), they classified the potential risk in the Arctic into the following: (1) operational, (2) regulatory, and (3) environmental. Also, in a study conducted by Dekel and Scotchmer (1990), they looked at the cost-sharing model used by the oil and gas companies. There is, however, an observed disturbing trend, which they deliberately spill oil to drive the cost up. There are questions on the morality of such an activity. This type of information would only discourage people from going ahead to buy insurance for their companies. The critique of such an immoral act is that even if the oil companies deliberately cause the oil spill, they will be covered by their insurance companies. This therefore creates a dilemma for insurance companies and regulators alike.

To address these problems, the US government proposed the 1990 Oil Pollution Act (OPA). In this law, the polluter is supposed to bear the cost of up to USD75 million. Nonetheless, the BP oil spill incident exposed the flaws of this law. This is because the USD75 million was woefully inadequate in dealing with such a problem (Harder, 2011). This was the case because we never anticipated that we would have an oil spill of such a magnitude. In the literature, a study was carried out, which compared the OPA and IMO regulations on oil spills. They concluded that both instruments are currently inadequate and lack the provisions to deal with certain types of oil spills like the BP oil spill (Schoenbaum, 2012). In addition, other researchers like Maniloff (2018) argued that OPA did not make the polluters pay enough for polluting. Soto-Oñate and Caballero (2017) also identified among other things the lack of transparency and consistency with laws, lack of proper guidelines to submit claims, and long wait time periods, the affected have to wait before the polluter takes responsibility. Understanding such deficiency, in this chapter, we hereby propose a hybrid model using Bayes' theory and the loss function to evaluate the oil spill insurance problem. This is necessary because to our best knowledge, hitherto, there are no existing tools that could objectively predict the insurance for Arctic-going ships. In addition, considering the dynamic nature of the Arctic using a hybrid Bayesian-loss function model would enable future researchers and practitioners to update the tool once new information becomes more available and accessible.

3 Methodology

As mentioned earlier, the method in this chapter adopts a combination of Baye's theory and the loss function. Both are of artificial intelligence (*AI*) tools. In the literature, Viaene et al. (2005) developed a Bayesian-based model to identify fraud claims for insurance, while Cowell et al. (2007) used the "OR" and "AND" gates in the Bayesian network for insurance provisions for businesses that run online. Fire insurance was investigated using a Bayesian-based model developed by Cheng and Hadjisophocleous (2009). In their study on developing insurance premiums, Li et al. (2010) leverage the use of Bayesian theory to develop a pricing system for premiums after a catastrophic event like the oil spill has occurred. Furthermore, to deal with the insurance of the so-called driverless cars, Sheehan et al. (2017) used the BN idea to develop a model for calculating the insurance premium for such cars. Using the hierarchical Bayesian model, Zheng et al. (2019) built a model to calculate the damage for insurance purposes during extreme crashes.

For the loss functions, many applications have been documented in the literature to assess loss. For example, Zadakbar et al. (2015) developed a loss function to assess the loss associated with the BP Texas City accident. Furthermore, Hashemi et al. (2015) developed a model based on the loss function for assessing loss in the processing industry. Abimbola and Khan (2018) presented a consequences model for evaluating a blowout incident for oil and gas production. The literature review on the application of the Bayesian network and the loss function shows that they have not been used together before and so this chapter is a pioneer in presenting such a platform.

3.1 Theory behind the Bayesian Network and loss function

The relation underlining the Bayesian network is shown in Eq. (1). Some literatures refer to it as the conditional probability equation, as follows:

$$P(A|B) = \frac{P(B|A)P(A)}{P(B)} \tag{1}$$

where A and B are referred to as the events, $P(A|B)$ is the probability that an event A will occur, conditioned on the fact that event B has happened already, and $P(B|A)$ is the probability of event B happening, conditioned on the fact that event A has happened already. However, the loss function concept is based on the Taguchi et al. (1989) concept of maximum loss. There are different types of loss functions that include squared error loss functions, absolute error loss function, quadratic loss function, inverted normal loss function, inverted beta loss function, inverted gamma loss function (Khan et al., 2016), to name but a few. Various types of loss functions are suited for different situations and scenarios. Also, loss functions are useful to estimate not only just the loss profiles, but also the risk profiles. Risk consists of threat, vulnerability, and the impact (Afenyo et al., 2022; Cox Jr, 2008). The concept proposed by Taguchi et al. (1989) can be expressed as follows:

$$L(z) = D(z - K)^2 \tag{2}$$

where D is a proportionality constant; K is the desired target value, and z is the parameter that we try to measure. However, one faces a lot of challenges in solving Eq. (2). Therefore, Spiring (1993), who tremendously addressed this shortcoming, referred to it as the inverted normal probability loss function (INLF), as follows:

$$L(z, K) = MAP\left\{1 - e^{-\frac{(z-K)^2}{2w^2}}\right\} \tag{3}$$

where MAP is the maximum value of parameter and w is the shape factor. Eqs. (4), (5) (see below) are other equations that are used in addition to Eqs. (1)–(3) to estimate the insurance premium. To calculate the claim from an affected party, Werner et al. (2016) proposed the following:

$$\text{Pure Premium} = \text{Claim frequency} \times \text{Claim Severity} \tag{4}$$

where Eq. (4) is further broken down and expanded into the following:

$$\text{Premium} = \text{Losses} + \text{LAE} + \text{UW Expenses} + \text{UW Profit} \tag{5}$$

where LAE is the loss adjustment expense, UWExpenses is the underwriting expense, UWProfit is the underwriting profit. To link the Bayesian network to calculate the insurance premium, we start with a characteristic vector $X = (X_1, X_2, \ldots, X_N)$. The constituents of the vector are known when the company decides to buy an insurance scheme. In perspective, if X_1 is the thickness of hull, X_2, the age of vessel, X_3 the type of hull, X_4, the expected number of days of the trip, and X_5 is the month of travel and we denote $i \in I$ as the accidents types that could lead to an oil spill; e.g., $i=1$ refers to sinking $i=2$ refers to allision, $i=3$ refers to grounding, $i=4$ refers to collision, etc. For each i, let $pi(X)$ be the conditional probability that accident i occurs during a trip identified by X. Furthermore, for each i, let $Fi(\cdot|X)$ be the conditional distribution of loss related to oil spill given that the accident has already occurred. This means that the probability that the loss has occurred, given that an accident i occurs, does not exceed y dollars, given by $Fi(y|X)$. On the other hand, the projected loss of trip X is described as $E[y] = X\, I \in I\, pi(X) \times Ei\, [y|X]$. Assuming that Z is a monetary amount deductible (e.g., dollars) and α is the risk loading, the premium can be described as follows:

$$P = (1 + \alpha) \times X\, i \in I\, pi(X) \times Ei\, [(y - Z) + |X] \tag{6}$$

where $(y - Z) + = y - Z$ if $y > Z$ and 0 otherwise. Assuming the insurance company decides to cover all losses due to oil spill incidents, the future losses, $Fi(y|X)$ could be described as follows:

$$\begin{aligned} y = {} & (\text{loss in year 1}) + (\text{loss in year 2})/(1 + r) \\ & + (\text{loss in year 3})/(1 + r)\, 2 + (\text{loss in year 4})/(1 + r)\, 2 \\ & + (\text{loss in year 5})/(1 + r)\, 2.... \end{aligned} \tag{7}$$

where r is the discounted cash flow.

4 An overview of industry experience in Arctic shipping insurance

To get an overview of the industrial experience on Arctic shipping insurance, in this study, we have conducted a panel discussion with four experts in 2021 who have worked with a real-world company that offers insurance to the maritime sector and covers financial protection and assists them in handling the disaster. As per requests from the experts, we have promised strict confidentiality and that the identity of the company would not be revealed in any published materials. Thus, for the rest of the chapter, this real-world marine insurance company will be described as "the company."

Despite the challenges associated with marine-related activities in the Arctic region, the company still considers a strong need for Arctic shipping insurance. The following factors contribute to the demand for marine shipping insurance in the Arctic:

- Its existing members want to trade to this area for fisheries, to transport cargo to and from and within the Arctic region.
- Since a large part of Norway is located within the Arctic region, and there is a need for shipping, there is a need for insurance.
- It has members that use the Northern Sea Routes from North to East and West.
- There is a demand for tourism, and extrapolation cruises are taking place in the Arctic region.

- It is engaged in ensuring the energy sector conducting business activities in the Arctic region.

Regarding the unique features of Arctic shipping insurance compared to other marine insurance, the company offers P&I insurance, i.e., third-party liability insurance, and the general principle is that this type of insurance is worldwide covered. It does not have specific rules for P&I. However, if it knows the specific route of their customer, this would be very helpful. In addition, if a shipping line trade between Svalbard and Greenland is known, the company, after assessing it, might impose different insurance rules.

For ship-specific (including autonomous ships) actuarially fair insurance premium, the company applies different methods of reaching the ship premium. However, ships are considered a part of the fleet and thus not priced on an individual basis. To determine the ship premium, the company analyzes the type of fleet the member has. For instance, the member might have bulk ships or tankers. The company assesses the risks related to the fleet and then offers better coverage. The difference in various regions within the Arctic also plays an important role in insurance (e.g., the risk is lower in the Barents Sea than in the North Sea in terms of weather conditions and environmental resources). For a detailed assessment of the fleet, the company uses different approaches that include the pricing method based on statistics like historical data, and other qualitative factors like experiences, trade area, and the performance of the insured member. These criteria are used for P&I in general terms. On the other hand, the insurance for the hull machinery is different, and there might be an additional premium for a specific area for certain time. For the hull machinery, the company operates the market at market price, which is a different approach and is more ship specific. If the vessel is operating in a large part of the Arctic region, it might be considered as high risk. However, some vessels have attributes that can mitigate the risks and consequently contribute to avoid a higher premium. For instance, if the vessel has an experienced crew, has ice class, and/or tugboats to assist in case of incident, etc., these could mitigate risk and hence a lower premium. However, the company is more concerned about the cruise ships that usually operate only once per year. Other characteristics include the fact that the crew is not experienced and is used to operating in, say, Mediterranean waters.

In this case, the company requests its members to provide information about particular voyages and to provide quotes that exceed the warranty limit to insure the hull machinery. So, if the vessel operates in the Arctic region, they would charge an additional premium for that vessel for that particular voyage. They assess it based on the vessel's capability, crew, and particular characteristics expected at the time of quote. Indeed, there is a possibility that there is zero additional premium at a specific route at one point of the year. At the same time, it might have a higher additional premium at the other time of the year when the weather condition is severe. In addition, the company evaluates whether the customer has made a proper risk assessment and presented their planning and precautionary measures during the assessment. If the customer can illustrate that it is aware of the risks, the underwriter trusts them. However, if the customers fail to illustrate the awareness of the potential risks (e.g., insisting that there would be no risk), the underwriter could become suspicious and thus more reluctant to insure them. The company also double-checked the risk assessment made by the customer. For this purpose, they have staff who can check the ice conditions, use various tools to check the ship's condition, check the class records, ask about crew competence, and use Automatic Identification System (AIS) data to see historical trade patterns. The staff of the company or a third party hired by the company could visit the vessel if they were seriously concerned about its conditions.

To carry out the assessments accurately and professionally, the company has employed an expert group of underwriters and other staff from different backgrounds. In addition, it utilizes other sources to improve the quality of the assessments. For instance, the Polar Code has been enhanced to explain the minimum requirements for vessels and crews to trade in the Arctic. The extended Polar Code requires enhanced Ice class crew qualifications. Moreover, the Nordic insurance plan has guidelines that clearly state the minimum criteria for members to comply with. Also, there are many industry guidelines that serve as a manual in directing how Arctic operations and business should be done. Usually, it knows the ships that trade in specific known routes in the Arctic. In some cases where ships go on expeditions in unknown areas, the company would conduct case-by-case assessments; plus, it is noted that wintertime is different from the summertime. Once the customer signs the policy, he/she is obliged to inform any alteration of risks to the company.

The detailed assessment criteria used by the company serve as guidance for their members to evaluate any new and risky voyages. Based on the feedback received from the company, the ship's internal management might decide to cancel the new voyage due to high risks. In this case, the company considers that Arctic shipping is not one uniform trade. However, there are several factors that make the Arctic unique for a marine insurer, where the claim amount becomes very expensive when it happens. They include

- Remoteness
- Rough and unpredictable conditions
- Unique vulnerable nature
- Lack of infrastructures
- Long distances
- Poor communications
- Darkness for extended period
- Timing of the year
- Less mapped area
- Lack of resources in case of an accident
- Lack of resources to open the Northwest Passage, the Northern Sea Route, and other Arctic routes.

Regarding what motivates marine insurers to ensure Arctic shipping despite a high likelihood of risks, the company does not consider the Arctic as a region with a high likelihood of risks, thanks to the fact that, hitherto, major accidents have yet to take place in the Arctic. Ships sail and trade in the Arctic area during the summertime when there is less risk due to ice. When bulk carriers and tankers sail through the NSR, they always have an ice breaker following them. However, when any incident happens then, it is severe. They have experience with severe claims and difficulty. Recently, a fishing vessel ran aground while fishing shrimps, and fortunately a helicopter managed to rescue the 14 seafarers onboard. There was a request to remove the wreck, and the company spent two summers removing it due to ice and poor weather conditions. Despite the likelihood of a severe accident, the company still provides insurance to its customers who want to trade in the Arctic region because higher risks come with higher prices. For the hull machinery, the risks are even higher in the Arctic region. Due to sea ice in the Arctic, vessels could face the following problems: propeller damage, the seawater could enter the cooling system and the engines could stop, the snow and ice can affect the stability of the ship, fog can affect the navigation, and then there are unchartered areas in the Arctic, crew members might get affected due to icy conditions. The combination of risks in certain Arctic areas is the Arctic's uniqueness. The company could increase the insurance premium in certain areas to solve such complexities of risks. A higher insurance premium is not an issue for the members compared to other costs they must bear in the Arctic region.

To convince insurance companies to insure the vessels operating in the Arctic region, an applicant must be able to illustrate that it is able to mitigate the higher costs and uncertainty in for its Arctic-going ships. In this regard, the company suggested taking the following measures:

- Better infrastructures and resources need to be developed.
- There are very limited salvage and rescue systems available at lower costs in the Arctic region. In other words, there is a large area in the Arctic region without any assistance. Having available salvage at acceptable prices and, more preferably, free with the intention to protect nature can make the situation better in the Arctic region. The insurer cannot provide these services, and so the governments need to be responsible for these.
- To improve the risk assessment quality in the Arctic region, the company recommends mapping the sea belt around the Arctic Sea. For instance, trolling vessels operating in the Arctic region cannot store and share the data, and they cannot record the data regarding conditions in the sea belt. Mapping of the sea belt can provide a better understanding of the territories.
- Geopolitical tensions in this area, for instance, the dispute regarding who owns the land masses, create a need for a better agreement for developing infrastructure.
- Current sanctions on Russia by several countries (notably NATO members), including several Arctic countries, worsen and further complicate the situation in rescue operations and payment.

5 Analysis of case study with the proposed model

The Hammerfest, Norway area is chosen for the illustration of the model (Fig. 2).

Furthermore, the output for the illustration is the loss and the risk, which are key for calculating the premium. Since the premium is area and environmental specific, the information obtained from the illustration, could then be used for the premium estimation using Eq. (6). The probabilities for the different types of accidents resulting in oil spill is shown in Table 1. The sample oil spill incident case is the *Exxon Valdez* oil spill. Since, as mentioned earlier, we have yet to suffer a major oil spill in the Arctic, we improvise using the *Exxon Valdez* oil spill in Hammerfest, Norway.

6 Analysis of the case

Fig. 3 illustrates the process in analyzing the case. The first step involves evaluating the literature for factors contributing to the oil spill. In addition, other factors have been identified as the factors companies consider when insuring ships going to the Arctic. These factors would be part of

Fig. 2 The location of Hammerfest, Norway. *(Source: http://www.maplandia.com/norway/hammerfest/.)*

Table 1 Probabilities for different oil spill scenarios on Baffin Island (BI), Nunavut, Canada.

Type of vessel	Accident type	Probability of spill per transit
Fuel Tanker	Collision	0.00000031
	Grounding	0.00000032
	Allision	0.00000060

(Source: Anon. (2019). Prepared for WWF. WWF Canada's final written submission for the Nunavut Impact Review Board's reconsideration of the Mary River Project Certificate for Baffinland's Mary River "Phase 2" Proposal.)

the qualitative model to be built. This activity is supported by contacting experts and other practitioners who work directly on oil spill-related problems. The information is used to develop the qualitative-based model shown in Fig. 4A and B. Fig. 4A is a more detailed version of Fig. 4B, of which it is presented particularly to the same group for structural validation. This involves making sure the logic behind the model is sound. This is then transformed into a quantitative-based model. To operationalize the model, the following needs to be done. The first step is the elicitation of the probabilities for the factors and probabilities for the conditional probability

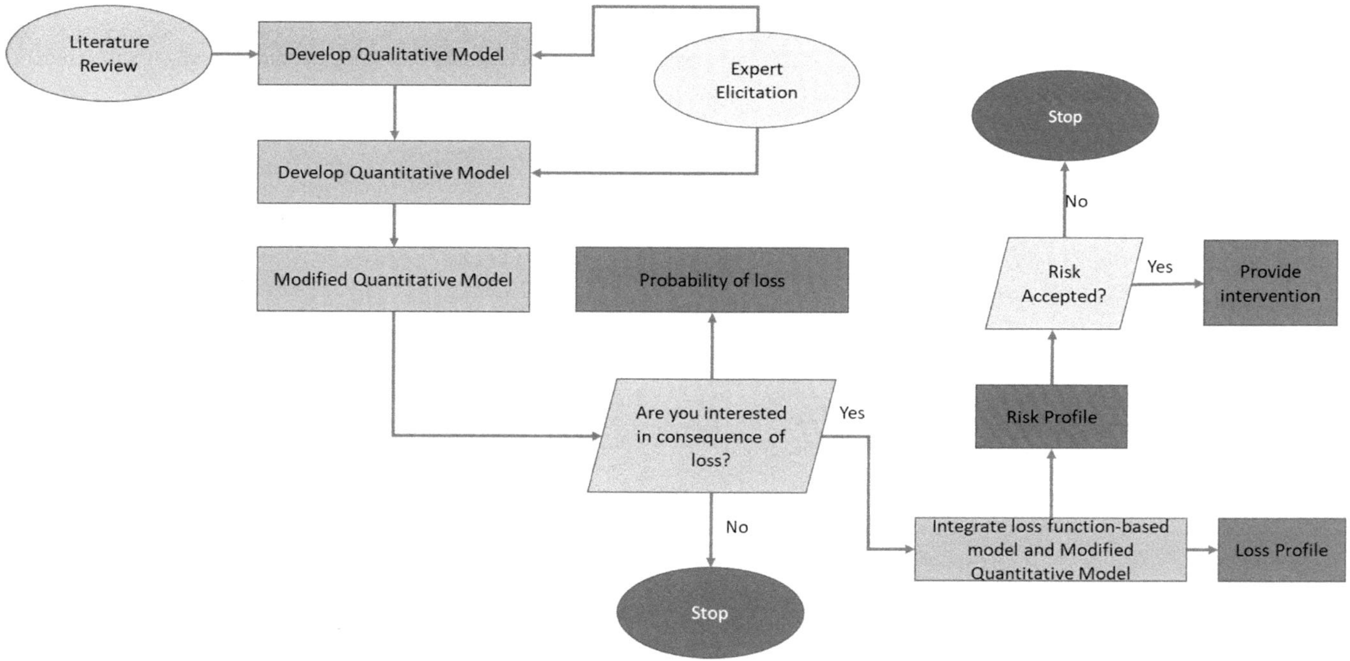

Fig. 3 The process of developing the model. *(Source: Authors.)*

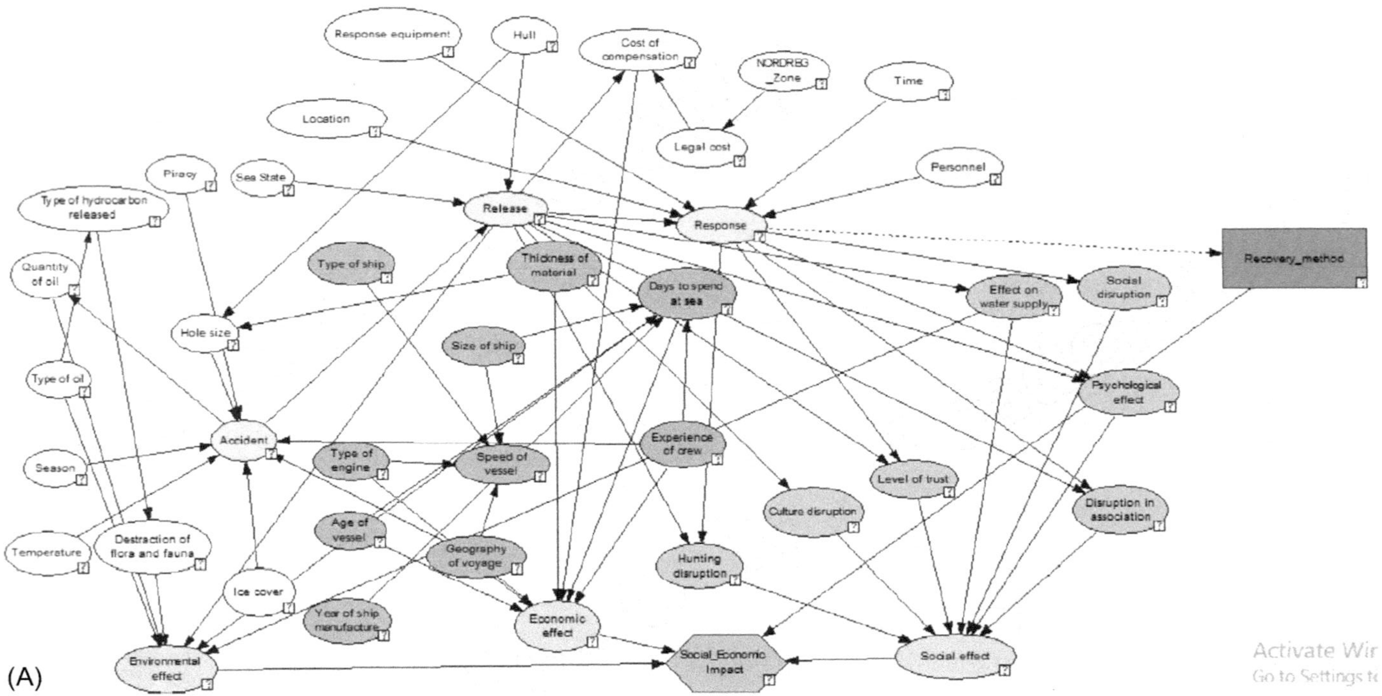

Fig. 4 (A) Qualitative model. (B) Qualitative model. *(A: Source: Afenyo, M., Ng, A. K. Y., & Jiang, C. (2022). A multi-period model for assessing the socio-economic impact of oil spills in the Arctic.* Risk Analysis: An international Journal, *42(3), 614–633. https://doi.org/10.1111/risa.13773. B: Source: Authors; Afenyo, M., Jiang, C., Ng, A.K.Y, & Lee, P.T. (2023). A Bayesian-loss function-based method in assessing loss caused by ship-source oil spills in the arctic area. Risk Analysis: An international Journal 43(8), 1557–1571. https://doi.org/10.1111/risa.14025.)*

(Continued)

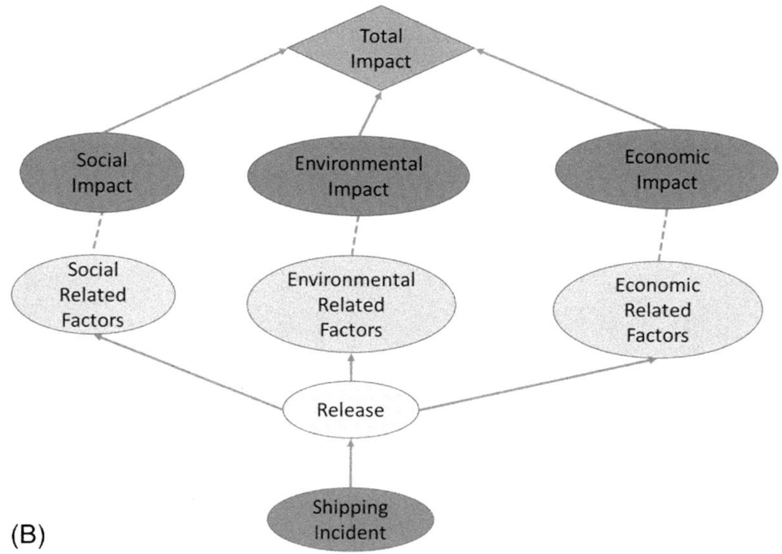
Total Impact
Social Impact
Environmental Impact
Economic Impact
Social Related Factors
Environmental Related Factors
Economic Related Factors
Release
Shipping Incident
(B)

Fig. 4, cont'd

tables (CPT). With the model, we proceed to assess the consequences using the loss functions. However, if we are not interested in the consequences evaluation, we could stop the process of operationalizing the Bayesian quantitative model. The integrated loss function-Bayesian model is then used to assess the risk of the voyage. The model also gives us the loss profile. If the level of risk predicted is too much, there is a need for some intervention; on the contrary, if the risk is okay, then the value is used to calculate the premium.

7 Results and discussions

Fig. 4A and B is the qualitative models (detailed and abridged). Figs. 5 and 6 are the loss and risk profiles. Summary of the detailed model in (A).

Fig. 5 shows the loss profile of the *Exxon Valdez* oil spill in Hammerfest, Norway. It shows the evolution of the amount of oil that is released in terms of US dollars. It gives us the possibility to deduce the amount of loss in monetary terms for any amount of oil spilled. It is an important trend to know, especially for maritime decision-makers. In the graph, the loss trend stabilizes around the maximum projected loss in the long run. This is approximately USD 370,000 and starts from a volume of 6000 m^3.

Similarly, the risk associated with the potential volume of oil to be released can be found in Fig. 6. It gives insurers the risk for different oil volume. Like the loss profile, the risk profile stabilizes around the maximum projected risk value, which for this illustration is approximately 12×10^{-5} and begin from a volume of 6000 m^3.

8 Conclusion

This chapter discusses the need for Arctic shipping insurance. To demonstrate the applicability of the model, it is applied to a hypothetical case in the region of Hammerfest in Norway. The outcome could be used in an insurance premium equation presented in this chapter for any fleet of vessels. Also, the model could serve as a blueprint for future works on Arctic shipping insurance. Understandably, the model is not without its challenges, and one of these is the development of the CPT. Furthermore, modifications would be required to apply to some autonomous ships. The current form is equally ok, but we might need more factors that are autonomous ship specific.

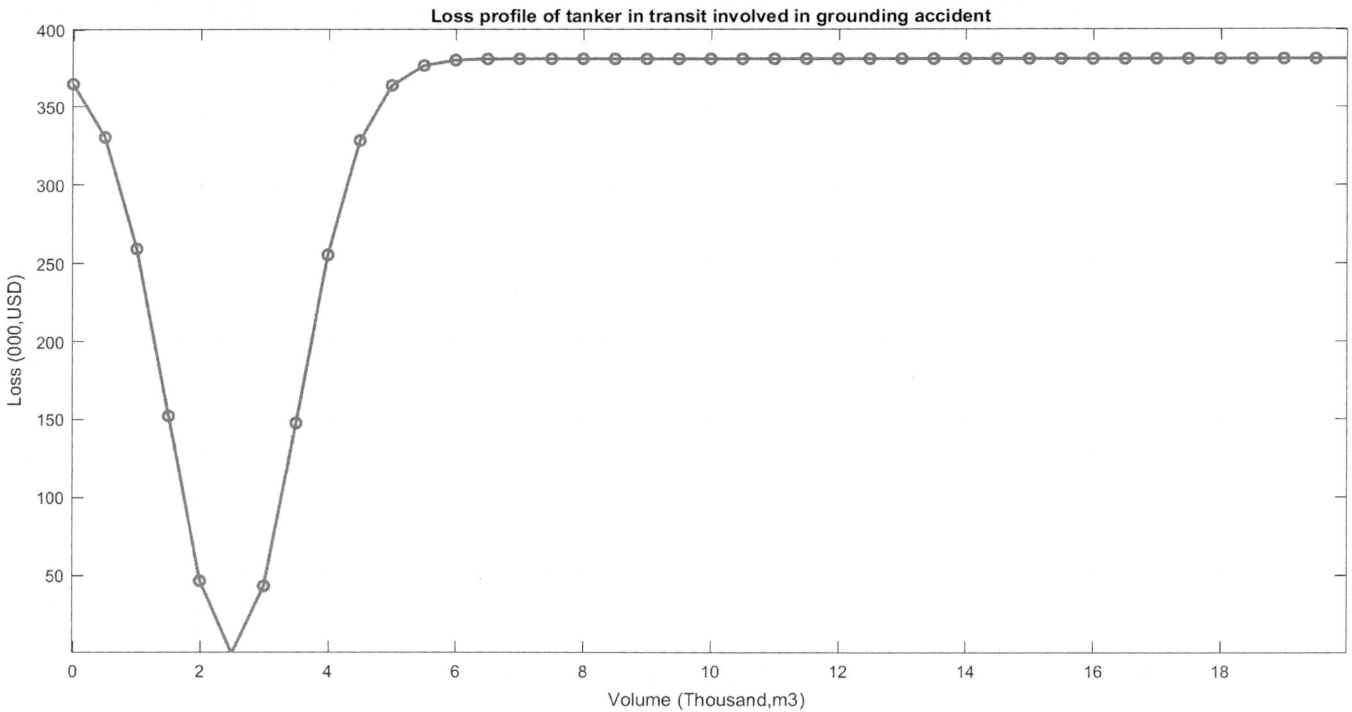

Fig. 5 Loss profile for grounding oil spill. *(Source: Authors.)*

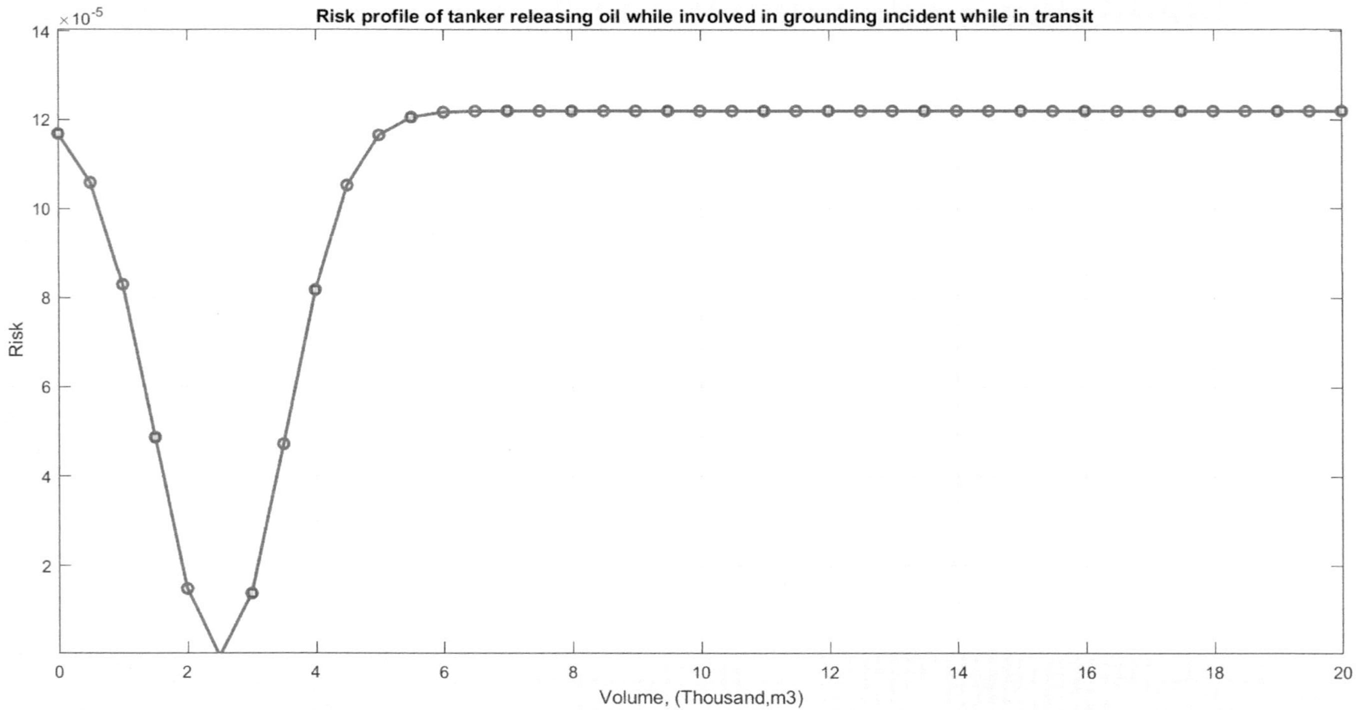

Fig. 6 Risk profile for fuel tanker grounding resulting in oil spill while making a transition. *(Source: Authors.)*

Acknowledgment

Part of this chapter is from Afenyo, M., Ng, A. K. Y., Jiang, C. (2022). A multi-period model for assessing the socio-economic impacts of oil spills during Arctic shipping. *Risk Analysis: An International Journal 42*(3), 614–633.

References

Abimbola, M., & Khan, F. (2018). Dynamic blowout risk analysis using loss functions. *Risk Analysis, 38*(2), 255–271.

Afenyo, M., Ng, A. K. Y., & Jiang, C. (2022). A multi-period model for assessing the socio-economic impact of oil spills in the Arctic. *Risk Analysis: An international Journal, 42*(3), 614–633. https://doi.org/10.1111/risa.13773.

Cheng, H., & Hadjisophocleous, G. (2009). The modeling of fire spread in buildings by Bayesian network. *Fire Safety Journal, 44*(6), 901–908.

Cowell, R. G., Verrall, R. J., & Yoon, Y. K. (2007). Modeling operational risk with Bayesian networks. *The Journal of Risk and Insurance, 74*(4), 795–827.

Cox, L. A. T., Jr. (2008). Some limitations of "risk = threat x vulnerability x consequence" for risk analysis of terrorist attacks. *Risk Analysis, 28*(6), 1749–1761.

DeCola, E. (2020). *Prepared for WWF. Compensation for ship-source oil spills: Considerations for modernizing Canada's marine liability regime to support indigenous communities.* Nuka Research and Planning Group, LLC.

Dekel, E., & Scotchmer, S. (1990). Collusion through insurance: Sharing the cost of oil spill cleanups. *American Economic Review, 80*(1), 249–252.

Dixon, R. (2021). While the world tore its hair out over the Suez, Russia saw an opportunity. *Washington Post.* https://www.washingtonpost.com/world/russia-suez-touts-arctic-sea-route/2021/03/29/576f6794-9097-11eb-aadc-af78701a30ca_story.html. (Accessed 27 April 2021).

Grabowski, M., Rizzo, C., & Graig, T. (2016). Data challenges in dynamic, large-scale resource allocation in remote regions. *Safety Science, 87*, 76–86.

Harder. (2011). Congress still grappling with oil spill liability. *National Journal, 1975.*

Hashemi, S. J., Ahmed, S., & Khan, F. (2015). Operational loss modelling for process facilities using multivariate loss functions. *Chemical Engineering Research & Design, 104*, 333–345.

Hinchey, G. (2017). *Arctic cruise ship owners ordered to pay $469K in costs for 2010 grounding. Canadian Broadcasting Corporation.* https://www.cbc.ca/news/canada/north/cruise-ship-clipper-adventurer-nunavut-judgement-1.3973937.

Johannsdottir, L., & Cook, D. (2019). Systemic risk of maritime-related oil spills viewed from an Arctic and insurance perspective. *Ocean & Coastal Management, 179*, 104853.

Khan, F., Wang, H., & Yang, M. (2016). Application of loss functions in process economic risk assessment. *Chemical Engineering Research & Design, 111*, 371–386.

Li, L., Wang, J., & Leung, H. (2010). Using spatial analysis and Bayesian network to model the vulnerability and make insurance pricing of catastrophic risk. *International Journal of Geographical Information Science, 24*(12), 1759–1784.

Maniloff. (2018). An estimate of the producer cost of liability for oil spills. *Applied Economics Letters, 25*(5), 348–353.

Panahi, R., Afenyo, M., Ng, A. K. Y., & Haeri, F. (2020). A novel approach in probabilistic quantification of risks within the context of maritime supply chain: The case of extreme weather events in the Arctic. *Accident Analysis & Prevention, 144*, 105673.

Russon, M. (2021). The cost of the Suez Canal blockage. *BBC News.* https://www.bbc.com/news/business-56559073. (Accessed 27 March 2021).

Schoenbaum, T. J. (2012). Liability for damages in oil spill accidents: Evaluating the USA and international law regimes in the light of Deepwater horizon. *Journal of Environmental Law, 24*, 3.

Sheehan, B., Murphy, F., Ryan, C., Mullins, M., & Liu, H. (2017). Semi-autonomous vehicle motor insurance: A Bayesian network risk transfer approach. *Transportation Research Part C, Emerging Technologies, 82*, 124–137.

Soto-Oñate, D., & Caballero, G. (2017). Oil spills, governance, and institutional performance: The 1992 regime of liability and compensation for oil pollution damage. *Journal of Cleaner Production, 166*, 299–311.

Spiring, F. A. (1993). The reflected normal loss function. *Canadian Journal of Statistics, 21*(3), 321–330.

Taguchi, G., Elsayed, E. D., & Hsiang, T. (1989). *Quality engineering in production systems*. McGraw-Hill.

Theocharis, D., Rodrigues, V. S., Pettit, S., & Haider, J. (2019). Feasibility of the Northern Sea route: The role of distance, fuel prices, ice breaking fees and ship size for the product tanker market. *Transportation Research Part E: Logistics and Transportation Review, 129*, 111–135. https://doi.org/10.1016/j.tre.2019.07.003.

Viaene, S., Dedene, G., & Derrig, R. A. (2005). Auto claim fraud detection using Bayesian learning neural networks. *Expert Systems with Applications, 29*(3), 653–666.

Wang, J., Tan, J., Guo, S., & Wang, S. (2018). High-value transportation disruption risk management: Shipment insurance with declared value. *Transportation Research Part E: Logistics and Transportation Review, 109*, 293–310. https://doi.org/10.1016/j.tre.2017.11.013.

Werner, G., Modlin, C., & Watson, W. T. (2016). *Basic ratemaking* (5th Edition). Casualty Actuarial Society.

Zadakbar, O., Khan, F., & Imtiaz, S. (2015). Development of economic consequence methodology for process risk analysis. *Risk Analysis: An International Journal, 35*(4), 713–731.

Zhen, X., Li, Y., Cai, G., & Shi, D. (2016). Transportation disruption risk management: Business interruption insurance and backup transportation. *Transportation Research Part E: Logistics and Transportation Review, 90*, 51–68. https://doi.org/10.1016/j.tre.2016.01.005.

Zheng, L., Sayed, T., & Essa, M. (2019). Bayesian hierarchical modeling of the non-stationary traffic conflict extremes for crash estimation. *Analytic Methods in Accident Research, 23*, 100100.

CHAPTER EIGHT

Evaluation of Arctic resources exploration, management, and supply chains

1 Introduction

For economic and human development, energy plays a critical role, and a significant part of the energy used for progress is obtained by extracting natural resources in the Arctic. Although the Arctic is rich in natural resources, a significant part of these resources is concentrated in the following three areas: the Beaufort Sea, the northwest of the Russian Arctic, and the Canadian Arctic Archipelago. According to a study conducted in 2020, the accessible oil and gas reserves for world economic development are left for only the next 50 years. Thus, exploring and managing Arctic natural resources is essential to overcome the deficiency in energy resources (For further details, see: https://arctic.review/future/natural-resources/). Due to climate change and ice melting, resource extraction activities have increased in the Arctic, which also caused an increase in demand for shipping. Because of challenging climate conditions and remote locations, resource exploration, and ship navigation also need support and assistance from third-party service providers.

In addition to resource exploration and cargo ship navigation, the melting ice has opened the door for tourism, another important source of income for the Arctic region. The Arctic has stunning scenery and plenty of opportunities for outdoor activities. Within a decade, from 2006 to 2016, there was a significant increase in summer and winter tourism. For instance, the demand for summer tourism has increased fourfold, and winter tourism by 600%. Thus, it is crucial to evaluate the impact of COVID-19 and to suggest suitable measures to overcome these impacts on the flourishing tourism industry in the Arctic region (For further details, see: https://wellbeingintl.org/tourism-in-the-arctic-a-catalyst-for-good-or-bad/).

Safe and Sustainable Arctic Shipping Management and Development
https://doi.org/10.1016/B978-0-443-18849-7.00008-7
Copyright © 2024 Elsevier Inc.
All rights are reserved, including those for text and data mining, AI training, and similar technologies.

The unique features of the Arctic region, like ice on water, severe weather, and remoteness, also impose additional risks to the Arctic supply chain compared to the same supply chain in other parts of the world. All these issues related to the link between resource extraction and shipping needs, the role of the third party in supporting the resource extraction and ship navigation, the resupply of tourism, and finally, the risk factors in supply chain disruption are explained and discussed in this chapter.

2 A link between the resource exploration trends and shipping needs

There is a close and significant link between resource exploration trends and shipping needs in the Arctic. Ice melting in the Arctic has increased access to larger marine areas and increased navigational seasons (Gunnarsson, 2021). Thus, ice melting is the main reason for linking resource extraction and shipping demand in the Arctic. The exploration of resources and several other associated activities, such as a rise in tourism, and social and economic progress, are pushing shipping trends upward (https://arctic-council.org/news/first-arctic-shipping-status-report-increase-shipping-traffic/; https://foreignpolicy.com/2020/10/13/arctic-competition-resources-governance-critical-minerals-shipping-climate-change-power-map/). The growth of shipping in the Arctic depends on efficient collaboration between the companies engaged in resource extraction and their shipping partners (Gunnarsson & Lasserre, 2023).

Demand for shipping has increased significantly in the Arctic and Northwest Passage. The vessel growth in the Arctic has risen by 25% from 2013 to 2019. In the Canadian Arctic, within 10 years (2009–2019), the number of vessels entering the Arctic Ocean was multiplied by 1.92, while in the Northern Sea Route (NSR), within 4 years (2016–2019), the number of ships multiplied by 1.58. Natural resource extraction is the main factor influencing the shipping growth in the Arctic. For instance, in the Canadian Arctic, the increase in shipping needs was due to the following reasons: fishing activities, which increased the demand by 106.2%; cargo handling activities, which increased the demand by 122%; general cargo increased the demand for shipping by 156.5%; and finally mining activities increased the demand for dry bulk vessels by the highest number that is by 288.9%. Mining extraction activities in several places in the Arctic and subarctic, for example, Voisey's Bay (Labrador), Raglan (Quebec), and Mary River (Baffin Island, Nunavut), have generated shipping needs. For instance,

Baffinland Iron Mines started its operation in 2015 in Mary River. They transferred 920,000 tons of ore via Milne Inlet port during the first year and later increased the demand for ships to carry 4.1 million and 5.1 million tons in 2017 and 2018, respectively. The company has now increased the demand for ships to transfer 12 million tons of ore annually. Also, the shipping needs have been realized to support the logistical activities of two gold mines north of Rankin Inlet. To meet the demands of mining operations, the Canadian transport company, Fednav operates specialized vessels capable of handling ice and severe weather and serves the two mines located in Deception Bay in northern Quebec (https://www.ispionline.it/en/publication/future-arctic-route-maritime-trade-geopolitical-revolution-32054#:~:text=It%20is%20apparent%20that%20the,passages%20in%20Canada%20and%20Russia).

In the Russian Arctic, companies extract resources throughout the year and continuously export these resources out of the Arctic despite the harsh weather and unfavorable working conditions, and this is possible because of the support from the Russian government. Also, resource exploration activities have increased the need for shipping and Russian presence in the Arctic. The Russian policy of converting Arctic gas to LNG and transferring oil and gas via Arctic terminals and ships instead of pipelines has significantly contributed to increasing the shipping demand and causing flexibility to share these resources with other destinations besides Europe (Gunnarsson & Lasserre, 2023).

In Russia, within four years (2016–2019), tanker vessels increased by 67.5%. There is a substantial growth in LNG vessels, which increased from zero to 507, and the number of ships with icebreaker equipment increased by 238%. The oil and gas activities in the following Arctic areas: Barents Sea (Prirazlomoye and Varandey Oil Terminals), the Yamal peninsula, and Ob Bay (Sabetta, Novy Port main terminals, and the Arctic LNG 2 terminal) have increased the tanker traffic. Oil production activities have also generated a significant part of shipping demands. For instance, Gazprom Neft, one of Russia's largest oil producers, increased its production at Novoportovskoye field in Yamal by 19% in 2018. Its export destinations are Norway, the UK, France, and the Netherlands. Also, there is the potential for an increase in bulk traffic in the Russian Arctic due to the opening of coal, lead, and zinc mines and the use of the port of Murmansk to transfer more ore (https://www.ispionline.it/en/publication/future-arctic-route-maritime-trade-geopolitical-revolution-32054#:~:text=It%20is%20apparent%20that%20the,passages%20in%20Canada%20and%20Russia;

https://www.highnorthnews.com/en/traffic-northern-sea-route-doubles-russia-aims-reduce-ice-class-requirements).

A significant part of the increase in shipping needs was from Novatek's Yamal LNG facility, which started its operation in 2017. Novatek's Yamal LNG is Russia's biggest natural gas producer and has exported more than 7 million tons of LNG to several destinations, including Norway, Western Europe, Brazil, and China. Another driving factor for generating shipping demands is the VostokCoal project, which is expected to produce about 30 million tons of coal annually from the Taybass basin on the Taymyr Peninsula in the coming years. The company will use ice-strengthened ships that can operate throughout the year to facilitate coal transfer (https://www.highnorthnews.com/en/traffic-northern-sea-route-doubles-russia-aims-reduce-ice-class-requirements).

The importance of Russia's strategic location in the Arctic is also evident from the NSR along the northern coast of Russia, which experienced increased demand for vessels. The number of vessel trips on NSR increased by 58% within 4 years, i.e., from 2016 to 2019, and a total of about 8329 sea voyages occurred on NSR in 4 years. The increased shipping demand in NSR was mainly due to the internal need to provide services to the ships and icebreakers and the increased demand for LNG vessels and tankers to deliver natural gas from the SW Kara Sea to European destinations. NSR is surrounded by four coastal seas: Kara Sea, Laptev Sea, East Siberian Sea, and Chukchi Sea. The Kara Sea experienced the highest demand of the four coastal seas due to two Russian hub ports—Murmansk and Arkhangelsk. In 2018, these two Russian ports handled about 50% of all traffic on the NSR and the main cargo consists of crude oil and metals. Moreover, Russia shared a significant portion (62%–73%) of all vessel operating companies on the NSR in these 4 years (Gunnarsson, 2021).

The Russia-Ukraine war and the world's section on Russia due to the war will decrease the resource extraction activities and shipping demands in the Russian Arctic in the short to medium run (Gunnarsson & Lasserre, 2023). However, despite the geopolitical tension and world sections, which resulted in a decrease in demand for Russian Arctic fossil fuels from Europe and the removal of foreign investment in the resource extraction projects in the Arctic region, the Russian government continues its interest in the area by announcing about 1.8 trillion rubles in 2022 for the development of NSR and the government continues its mega-projects on oil extractions, and started to collaborate with India and China in the Arctic region (https://www.csis.org/analysis/russian-arctic-threat-consequences-ukraine-war; https://worldcrunch.com/war-in-ukraine/russia-in-the-arctic). Since China is the leading manufacturer in the world, it is expected that the demand for the metals extracted from the Russian Arctic and used in the manufacturing of several products by China will remain high.

3 The role of third-party services in supporting resource exploration and shipping in the Arctic

Due to severe weather and unfavorable conditions for operations, resource exploration, extraction, and ship navigation in the Arctic are challenging tasks that are only possible to carry out with the help of third service providers. A vital stakeholder is the logistical service providers. For instance, GAC Norway As (GAC) developed a collaboration with Polar Logistics Group ApS (POLOG) to support the oil and gas sector in the Arctic. GAC has assisted the oil firms in northern Norway since 2007, when Hammerfest, Norway's LNG site, started working. The company also offers services in other areas of the Arctic, including the Barents and Kara seas. The collaboration with POLOG would be helpful to use its extensive network in providing logistic services to oil companies, such as ice runways and mountain-top camps. In addition to oil companies, GAC supported shipping companies navigating the Arctic. For instance, the shipping company OB River was the first to use the NSR to export LNG from Norway to Japan, and GAC assisted the OB River as their agent during this voyage (https://shipandbunker.com/news/world/681757-firms-form-partnership-for-arctic-oil-and-gas-services). Offshore oil and gas extractions involve several stakeholders who support the functions, such as the oil company as a leading operator, the rig owner and subcontractors, supply base operators, and transport service providers to move the cargo. The platform service vessels also assist the oil operating companies. They perform the dual roles of transportation and sometimes as storage capacity for the rig vessels (Borch, 2018).

Resource extractive activities in the Arctic negatively impact the environment and ecosystem. Third-party organizations and experts like research institutes and academics can assist the extractive industries and policymakers in mitigating these negative externalities. For instance, Overland et al. (2021) developed the Arctic Environmental Responsibility Index (AERI), which covers seven Arctic nations (the United States, Canada, Greenland, Finland, Norway, Russia, and Sweden) and 120 resource extractive companies. The index was developed by distributing a survey to 173 experts who are International Panel on Arctic Environmental Responsibility (IPAER) members. The index can serve as an informal governance mechanism to monitor and control the environmental performance of the companies involved in resource-extractive activities in the Arctic region. For instance, the results revealed that the following companies followed the best practices to protect the environment during extraction: Equinor, Total, Aker BP, ConocoPhillips, and BP. The following companies did not get a high ranking and are considered as the ones not contributing to protecting the environment: Dalmorneftegeophysica, Zarubejneft, ERIELL, First Ore-Mining Company, and Stroygaz Consulting.

However, some stakeholders have supported the efforts to mitigate the Arctic's resource extraction activities and shipping navigation. For instance, the Obama's administration prohibited using about 10 million acres for resource extraction, which postponed oil and gas drilling in the US Chukchi Sea (https://oceanconservancy.org/protecting-the-arctic/). Similarly, environmental groups are convincing shipping companies to avoid the Arctic route to protect the environment. Because of their efforts, three top shipping companies worldwide, including Kuehne + Nagel and Hapag-Lloyd, who own one-third of the market share of the shipping industry and 1366 vessels, voluntarily committed not to use the Arctic route (https://www.cbc.ca/news/canada/north/arctic-shipping-pledge-1.5424250).

Specific characteristics such as stormy ocean waves, wind, ice, and fog make the Arctic environment harsh and remote, creating hurdles for the resource extractive companies during drilling, causing damage to the infrastructure offshore (Adumene & Ikue-John, 2022) and onshore, and making the shipping navigation risky and challenging. Insurance companies play a significant role in mitigating risks for shipping companies by sharing important information and promoting effective strategies to avoid risks and ensure safe vessel navigation. For instance, shipping and insurance companies can utilize the information related to the latest incidents, routes, and ice maps on the Joint Hull Committee website. Similarly, insurance companies also provide reasonable insurance to the energy sector to cover the following: loss and damage to persons and property of the company as well as of the third party, shifting the wreck and evacuation expenditures, delay in operation activities, and loss of profit, and risks associated with construction and drilling (http://www.arctis-search.com/On+Insurance+and+Managing+Risk+in+the+Arctic).

Oil and gas drilling and vessel navigation in the Arctic require specialized equipment, technology, and vessels to protect the ecosystem and avoid accidents and disasters. For instance, third-party service provider ABB considered a world-class technology leader in electrification and automation, supported the energy shipping company Sovcomflot in operating the giant ice-class vessel. The fleet uses less fuel and is equipped with ice-breaking technology, facilitating navigation in up to 2.1 m of snow. ABB also provides digital connectivity for shipping companies operating in the Arctic with its Ability Collaborative Operation Centers throughout the world. More than 1000 vessels are connected and use the services of the Collaborative Operation Centres. ABB specialists evaluate operational shipboard systems in these centers, communicate equipment problems and issues, and guide maintenance services. Moreover, to support and guide the ships operating through the most risky and challenging route via the Arctic, ABB has developed a specific Marine Service Centre in the northern Russian port city of Murmansk, which provides nonstop services (https://new.abb.com/news/detail/58692/how-abb-technology-is-supporting-a-sustainable-future-for-arctic-shipping).

4 The resupply of tourism in the Arctic during COVID-19

The Arctic has a rich tourism history that dates back to 1836 when John Murray wrote the *Handbook for Travelers* which guided and encouraged cruise tourism. The first cruise ship started its voyage in the Arctic in 1869 (Remer & Liu, 2022), and later, the expedition of an American, Frederick Albert Cook, in 1908 to the North Pole attracted the attention of tourists in the region (Barik et al., 2022). Arctic tourism is unique and different from tourism in the rest of the world and is known as *Special Interest Tourism* (Maher et al., 2014) and *Polar Tourism* (Demiroglu & Hall, 2020). The distinctive features of Arctic tourism are its flora and fauna, and it is the favorite destination for tourists who love and like to explore nature. The Arctic region offers separate activities for tourists during summer and winter. For instance, in summer, tourists can enjoy hot air ballooning, private boat trips to explore mountains, icebergs, glaciers, fjords, and polar bears, participate in travel group activities like exploring the Indigenous people's culture and their lifestyle, joining the local games like sled dog adventures and staying at igloo. The main winter activity, also considered one of the main driving forces of tourism, is experiencing the Northern Lights in the Arctic region (Barik et al., 2022).

However, despite the old history and tourists' interest in the Arctic region, the tourism industry has yet to flourish due to ice and safety issues. The main hurdle was the lack of specialized cruise ships that could navigate through the ice and handle the harsh conditions of the Arctic. However, the situation changed a bit after the collapse of the USSR in December 1991 due to the use of ice-breaking ships for tourism purposes, which were previously reserved for military purposes (Remer & Liu, 2022). The use of ice-breaking cruise vessels resulted in growth in tourism in the Arctic.

COVID-19 has significantly affected tourism across the Arctic. During the first quarter of 2020, tourism faced a relatively low decline, with about a 50% reduction in the tourist arrival rate, compared to a drastic hit on the tourism industry in the second quarter of the same year due to the lockdown in the region to control the pandemic spread. For instance, several fully occupied hotels in 2019 experienced a sharp decline, i.e., from 100% occupancy to 10%–15% in 2020. Arctic tourism is divided into two seasons: summer, which is relatively short (about 2 months only) but regarded as the peak time for the tourism demand, and winter, which is long (about 9 months) and involves the most expensive activities such as watching the Northern lights. The summer tourist destinations in the Arctic are only Alaska and Norway, which were affected the most by COVID-19 because of the lockdown and decreased demand for international tourists. Compared to summer, the long winter seasons could accommodate several measures related to COVID-19 prevention, such as 14 days of quarantine, a COVID-19 vaccination certificate, and keeping social distance. Despite such, the demand for international tourists in winter was still low because most of the travelers came from Asia, North America, and Europe, which were still experiencing the second wave of COVID-19 and consequently experiencing the lockdown situation and an economic decline (Barik et al., 2022).

However, tourism development varies in different parts of the Arctic. For instance, some touristic destinations witnessed a decline in demand for tourism mainly due to their dependence on international tourists. In contrast, other parts experienced overtourism due to increased tourism demand from the local tourists who took the opportunity to explore outdoor recreation close to their hometown (Jóhannesson et al., 2022).

Several factors contribute to formulating the strategies to overcome the tourism crisis due to COVID-19, such as domestic versus international tourism, support from the government, and selection of the lifestyle of the tourists. The extreme volatility in demand for tourism, especially during the beginning of COVID-19, raised concern about cruise ship capacity. Also, the pandemic raises awareness about tourism's negative impact on climate and biodiversity in the Arctic. Although overtourism has some negative consequences, like harming nature and a decrease in the quality of life of local citizens of tourist destinations, COVID-19 has diverted the attention of policymakers and researchers from over-tourism to how to re-start tourism with COVID-19. The re-supply of tourism in the Arctic during COVID-19 requires the following points to be considered: the policymakers should differentiate between the international tourism market across the Arctic and domestic tourism within the Arctic and formulate different strategies for these two segments of tourism. The significant increase in domestic tourism during COVID-19 also identified the potential rich growth in demand available with domestic tourism. Thus, the tourism sector needs to understand and meet the demands of their local customers and integrate local context into their future Arctic policies. The variations in demand for international and domestic tourism also show no "one" solution for all tourist attractions in the Arctic.

Different strategies for local and international customers should be formulated by considering the geography (distance) and other factors like the role and impact of tourism on the local economy and ecosystem (Jóhannesson et al., 2022).

5 The risk of the Arctic supply chain disruption

The research has identified four areas of the Arctic supply chain. The first area covers the environmental and social aspects of spreading the Arctic supply chain. The second area examines the potential benefits of utilizing the Arctic Sea routes for fisheries and cargo movement, including container, liquid, and bulk. The third area of research estimates the navigation in the sea ice. The fourth and relatively new stream of line focuses on improving the reliability of the Arctic supply chain. The economic benefits associated with the extraction of natural resources and the use of NSRs have also attracted the attention of researchers to explore and check the reliability of the Arctic supply chains (Taarup-Esbensen & Gudmestad, 2022).

In general, risk factors in a supply chain can disrupt the smooth movement of goods, resources, money, and information and, consequently, delay manufacturing and delivery to the final customers (Zsidisin & Henke, 2019). Compared to a global supply chain in general, the Arctic supply chains are more complicated and unreliable from a technical and organizational perspective. Often, voyages are registered and mentioned on paper only because the actual voyage was either delayed or canceled due to unpredictable weather and ice and navigation difficulties (Marken, Ehlers, & Khan, 2015). The Arctic supply chain has more risk factors than the same sector's supply chain in other regions because it is exposed to two groups of risks. Any supply chain could encounter the first group of risks. In contrast, the second category consists of the risks specific to the Arctic region due to its climate and remoteness (Gunnarsson & Lasserre, 2023; Taarup-Esbensen & Gudmestad, 2022).

Taarup-Esbensen and Gudmestad (2022) divided the risks specific to the Arctic supply chain into four groups:

(1) Technical factors include the impacts of severe weather (for instance, polar lows, heavy winds, and snowfall), darkness, and sea ice on operational efficiency and navigation smoothness. Disruptions in electricity due to mechanical faults, inability to correctly predict the weather, lack of access to quality satellite coverage, poor infrastructure, and lack of efficient rescue and logistic services.

(2) Safety factors disrupting the Arctic supply chain include harsh weather conditions, lack of advanced communication technologies, health and safety of the crew members, lack of skilled workers familiar with the Arctic environment, language hurdles, and unexpected events like tsunamis and COVID-19.

(3) Environmental risk factors are adverse environmental effects due to oil spills and the nonavailability of equipment to handle an emergency.

(4) Reputational risks include factors such as not following the rules and regulations implemented by authorities, negative influence on the sea life due to pollution, and measures to handle any disaster that do not fulfill the criteria and expectations of stakeholders.

The above-mentioned risk factors increase the probability of accidents and disturbances in the Arctic supply chain.

Researchers recommend several strategies to mitigate the Arctic supply chain disruption risks. For instance, like Russian oil and gas companies, other resource extractive companies operating in the Arctic could control the whole supply chain and its activities, such as resource exploration, infrastructure development, production, and distribution of goods and services. This strategy is facilitated by improving collaboration and communication with different supply chain segments. Another good plan is to align their supply chain goals with government policies to receive government support in risk-mitigating efforts (Gunnarsson & Lasserre, 2023). Another strategy is applying advanced techniques and innovative methods to mitigate supply chain risks, such as using ice-breaking equipment and vessels and improving the port infrastructure (Dalaklis et al., 2018; Taarup-Esbensen & Gudmestad, 2022). As explained in the previous section, integrating third-party service providers to support resource extraction and ship navigation is an excellent strategy to avoid the hazard.

Despite following the impressive strategy to control the whole supply chain by resource extractive companies, the Russian Arctic supply chains face challenges due to the Russia–Ukraine war. The war has negative consequences across Russia. However, the Arctic supply chain is more vulnerable to world sanctions due to its fragile situation, mainly because of less developed infrastructure and massive dependence on advanced techniques and equipment due to ice. The world sections have negatively affected the transport and logistic segments of the Russian Arctic supply chain, especially logistics related to the export and import of the products. The Russian oil and gas industry largely depends on international shipping lines to import heavy manufactured parts from abroad and export raw materials from the Arctic. This disruption stresses the need to find alternative logistic strategies and alternative partners for their supply chain (https://thebarentsobserver.com/en/arctic/2022/04/transport-crisis-looms-russian-arctic). All eight Arctic states, including Russia, are members of the Arctic Council—the main governance channel for the Arctic region. The Arctic Council could not implement its policies and decisions without the agreement of all eight member countries. After the Ukraine war, the seven members of the Arctic Council jointly refused to cooperate with Russia, resulting in the suspension of joint efforts of the member states in all areas of the Arctic development, including research, environmental control, and formulating policies and strategies for shipping and resource extraction and development. Thus, the Ukraine war also boosted the new and potential risks threatening the reliability of the Arctic supply chain (https://www.swissre.com/institute/research/sonar/sonar2023/new-risks-in-arctic.html#:~:text=Potential%20for%20business%20disruptions%20in,global%20supply%20chains%20is%20elevated.&text=Arctic%20opening%20and%20permafrost%20melt,could%20trigger%20casualty%20insurance%20policies).

6 Conclusion

In the Arctic region, the main driving factor for increased shipping is the extraction of natural resources, while transit traffic remains marginal or even declines. This close and significant link between the growth of the two sectors (natural resources and ship demand) has been successful due to close collaboration between the two segments of the Arctic supply chain. Among the natural resources, mining activities, including iron ore and gold, have pushed the need for shipping. In Russia, the companies have been involved in resource extraction activities for the whole year mainly because of generous support from the Russian government. The Russian strategies of converting Arctic gas to LNG and using the Arctic terminals and ships instead of relying on pipelines for export have also accelerated the shipping demand.

The challenging work and navigation environment in the Arctic also makes it necessary to integrate the support of third-party service providers in resource exploration and ship navigation. Several stakeholders offer their services to extractive and shipping industries in the Arctic. One is logistic service providers, who provide services to the extractive sector, such as ice runways, mountain-top camps, transport, and storage facilities. The logistic service providers support ship navigation by serving as their agents. Research and academic institutes also assist the extractive industry, for instance, by providing environmental guidelines and frameworks to mitigate the negative impact on the environment and ecosystem. Insurance companies are an integral source of support for resource exploration and ship navigation, which is facilitated through information sharing and providing insurance to the person, property, and damage to a third party. Also, third-party service providers ease the difficult drilling and navigation operations by providing advanced equipment, technology, and digital connectivity.

Interestingly, COVID-19 has affected the Arctic tourism. It has positively accelerated the demand for domestic tourism. At the same time, it has declined international customers' demand, suggesting that policymakers should not follow one solution for Arctic tourism. Instead, the policymakers and tourism industry should integrate the demands and preferences of local tourists to utilize the potential growth of domestic tourism. Finally, to reduce the risks of the Arctic supply chain disruption, it is recommended that the extractive industry tries to control the whole supply chain to make it more efficient and reliable.

References

Adumene, S., & Ikue-John, H. (2022). Offshore system safety and operational challenges in harsh Arctic operations. *Journal of Safety Science and Resilience*, *3*(2), 153–168.

Barik, J., Pal, I., Ghosh, T., Mukherjee, S., & Mukhopadhyay, A. (2022). COVID-19 pandemic in the Arctic and subarctic. In *Pandemic risk, response, and resilience* (pp. 143–156). Elsevier.

Borch, O. J. (2018). *Offshore service vessels in high artctic oil and gas field logistics operations: fleet configuration and the functional demands of cargo supply and emergency response vessels*. Available at https://core.ac.uk/download/pdf/225907014.pdf.

Dalaklis, D., Drewniak, M. L., & Schröder-Hinrichs, J. U. (2018). Shipping operations support in the "high north": Examining availability of icebreakers along the Northern Sea route. *WMU Journal of Maritime Affairs*, *17*, 129–147.

Demiroglu, O. C., & Hall, C. M. (2020). Geobibliography and bibliometric networks of polar tourism and climate change research. *Atmosphere*, *11*(5), 498.

Gunnarsson, B. (2021). Recent ship traffic and developing shipping trends on the Northern Sea route—Policy implications for future Arctic shipping. *Marine Policy*, *124*, 104369.

Gunnarsson, B., & Lasserre, F. (2023). Supply chain control and strategies to reduce operational risk in Russian extractive industries along the Northern Sea route. *Arctic Review on Law and Politics*, *14*, 21–45.

Jóhannesson, G. T., Welling, J., Müller, D. K., Lundmark, L., Nilsson, R. O., de la Barre, S., et al. (2022). *Arctic tourism in times of change*. Nordic Council of Ministers.

Maher, P. T., Gelter, H., Hillmer-Pegram, K., Hovgaard, G., Hull, J., Jóhannesson, G. T., et al. (2014). Arctic tourism: Realities and possibilities. In *2014. Arctic yearbook* (pp. 290–306).

Marken, V. B., Ehlers, S., & Khan, F. (2015). Delay risk analysis of ship sailing the northern sea route. *Ship Technology Research*, *62*(1), 26–35. https://doi.org/10.1179/0937725515Z.0000000007.

Overland, I., Bourmistrov, A., Dale, B., Irlbacher-Fox, S., Juraev, J., Podgaiskii, E., et al. (2021). The Arctic environmental responsibility index: A method to rank heterogenous extractive industry companies for governance purposes. *Business Strategy and the Environment*, *30*(4), 1623–1643.

Remer, M., & Liu, J. (2022). International tourism in the Arctic under COVID-19: A telecoupling analysis of Iceland. *Sustainability*, *14*, 15237. https://doi.org/10.3390/su142215237.

Taarup-Esbensen, J., & Gudmestad, O. T. (2022). Arctic supply chain reliability in Baffin Bay and Greenland. *Polar Geography*, *45*(2), 77–100.

Zsidisin, G. A., & Henke, M. (Eds.). (2019). *Revisiting supply chain risk*. Cham: Springer.

SECTION D

Environmental and social impacts of Arctic shipping

CHAPTER NINE

Environmental and socio-economic implications of Arctic development

1 Introduction

As discussed in previous chapters, there are many consequences associated with Arctic shipping and related operations. However, an area that we, so far, have not really discussed much is the social and environmental implications, especially for carbon emissions. During shipping as well as other operations in the Arctic carbon dioxide is emitted just like it happens in open waters. This has become a major concern for environmentalists as well as most governments and institutions concerned about climate change. Therefore, there is a major discussion about the type of fuel that ships should be using for navigation. This is particularly disturbing if we look at the Arctic in a global context where efforts are being made to make sure the concept of sustainability is pushed in every activity that we engage in. To put things in perspective, the shipping industry alone uses 5% of the global oil production, this is a high number if we consider the size of the industry in comparison with others like the Tech and the healthcare space.

According to the International Maritime Organizations (IMO), the shipping industry releases approximately 3% of the world's Carbon dioxide equating to about 1.05 billion tons (IMO, 2020). The current fuel used by most ships is heavy fuel oil (HFO). This fuel contains Sulfur, Nitrogen, and metals in addition to aromatics, paraffins, olefins, and asphaltenes. The compositions present very devastating consequences for the environment. There is a huge advocate for the use of liquified natural gas (LNG) powered ships. While LNG is cleaner and better, there is the challenge of cost involved in terms of initial investments. In addition, the percentage of

Safe and Sustainable Arctic Shipping Management and Development
https://doi.org/10.1016/B978-0-443-18849-7.00005-1
Copyright © 2024 Elsevier Inc.
All rights are reserved, including those for text and data mining, AI training, and similar technologies.

Methane which is one of the ingredients responsible for global warming is also high hence the continued search for a cleaner alternative (Jacoby, 2022). The use of methanol, hydrogen, and fuel cells remain the other alternatives. Each of them has its challenges but is environmentally friendlier than the current HFO and LNG to a large extent (Atilhan et al., 2021; Svanberg et al., 2018). For example, green hydrogen which is the most environmentally friendliest among them needs to be technically viable in terms of production, economics as well as safety bit of it (Atilhan et al., 2021). In some cases, these requirements make investors and other interested companies a bit skeptical about breaking even as soon as possible. The question they keep asking is what if the investment never enables them to make a profit, they make it a huge gamble, and becomes uncharted territory.

In terms of the social impacts, Afenyo et al. (2021) and Afenyo, Ng, and Jiang (2019, 2023) have highlighted the social implications of Arctic shipping. In their study on oil spills in the Arctic areas, they argued that an oil spill in Arctic waters might impact the livelihood of the population that depends on the Arctic ecosystem for survival. Losing one's source of livelihood can have a serious effect on the family makeup and by extension on the community. People move to where they can find jobs. If they do not possess the skills to secure alternative employment, they might resort to criminal activities to make ends meet. In some cases, this could result in breaking down the family structure, where spouses might split, and children might move from the parents. Therefore, the consequences of Arctic shipping operations go beyond the economic and environmental aspects. In some cases, even the way of worship could be affected. According to some of our discussions with the local population in northern Canada, some local communities worship animals, the ocean, and flora species in the Arctic. Destruction or any negative effect on these could pose serious psychological impacts on these people and communities (Afenyo et al., 2021).

2 Environmental impacts

The environmental impact will be assessed in two ways. The first is the release of oil from accidental spills, and the second is carbon emissions from ships. Fig. 1 is a model developed by Afenyo et al. (2021) and captures some of the factors responsible for the economic, social, and environmental impacts. In terms of environmental impacts, some of the factors include the type of hydrocarbons released, the quantity of oil released, the state of the sea at the time of oil released, the temperature, and if ice is present or not.

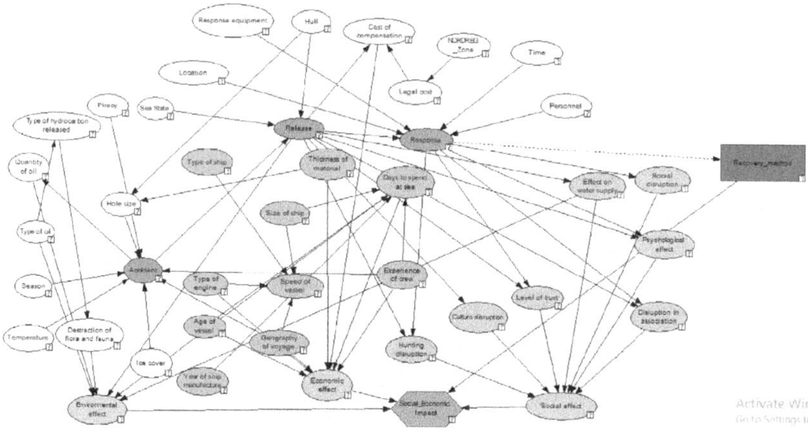

Fig. 1 Socio-economic impact model developed by Afenyo et al. (2021).

These together would affect how much impact it will have on the flora and fauna in the area. Should the food that the indigenous people will feed on become contaminated it can result in carcinogenic diseases. Should the oil get encapsulated in ice, it is most likely to be transported miles from where the original release takes place and once the temperatures increase, the oil gets released into the ocean and further pollutes that region too and potentially causes health problems for people eating fish and other seafood from that area. Chapter 6 is mainly dedicated to oil spills in the Arctic and so will not be repeated in this chapter. On the other hand, the development and application of Fig. 1 has been, to some extent, discussed in Chapter 7.

For the carbon argument and implication of Arctic shipping, the next paragraphs are dedicated to how the emissions are calculated and what would be the benefits of using Arctic Shipping. Basically, this shows that, even though the impacts may remain in terms of the release of carbon, it will nonetheless cut emissions tremendously. This can be demonstrated through an analysis conducted by us on the shipping of grain from Churchill to Europe. The analysis is aimed at presenting the carbon benefit of using the Churchill (Canada) Port-Mersin (Turkey) shipping route as opposed to using the Vancouver (Canada)-Mersin (Turkey) shipping route. Further, the carbon benefit is compared to other activities including flight, rail, and driving a car. Two methods of analysis are presented mainly to show the advantages and disadvantages of the two methods. The first method is the activity-based carbon emission, and the second is the energy-based carbon emission calculation. The idea of these calculations is to show the level of carbon benefits of using the Arctic as opposed to using the traditional routes and other means of transport. That is not to downplay the large impact of carbon emissions while shipping. However, the argument here is that using Arctic shipping would help reduce the percentage of carbon emissions and by extension reduce global warming. To demonstrate such, the following equation is applied where

$$CO_2\text{Emissions} = \text{Transort Volume} \times \text{average transport distance} \times CO_2\text{emissions factor per tonne} - \text{km} \quad (1)$$

The transportation of grains is used as a case study in this example. Tables 1 and 2 show the inputs for the two routes under comparison.

The results of the calculations are presented in Fig. 2. The results illustrate that the Churchill to Mersin route has a carbon footprint of 3907.2 tons

Table 1 Values of variables for Churchill to Mersin.

Volume of grain (tons)	Average transport distance (Churchill to Mersin) (km)	CO_2emission factor (gCO_2/ton-kms)
35,000	13,289.95	8.4

Table 2 Values of variables for Vancouver to Mersin.

Volume of grain (tons)	Average transport distance (Vancouver to Mersin) (km)	CO_2emission factor (gCO_2/ton-kms)
35,000	22,816.64	8.4

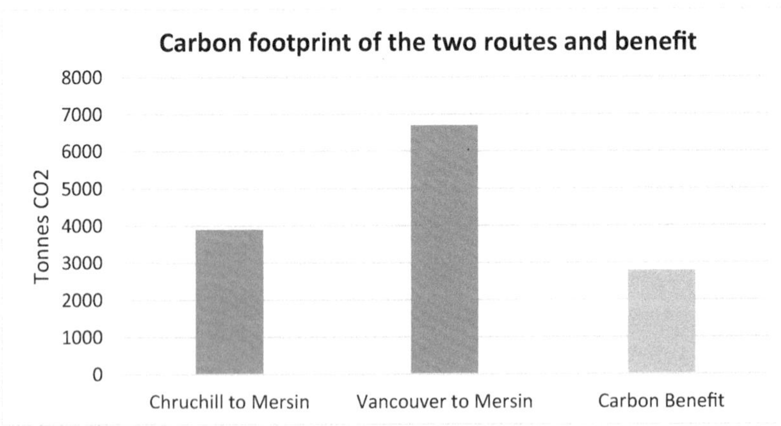

Fig. 2 The carbon footprints of the two routes and the associated carbon benefit.

CO_2, while the Vancouver-Mersin route produces 6708.1 tons CO_2. This translates into a carbon benefit of 2800.8 in favor of using the Churchill-Mersin route to transport grains. In other words, a lot of CO_2 will be reduced if this route is used. To put this in perspective, an average American releases 23 tons of CO_2 per year meaning that using the Churchill route will help eliminate the Carbon emission of 120 Americans in a year.

Furthermore, we need to consider the peak season where, for example, 20 ships are anticipated, meaning a total volume of grain transport to be 700,000 tons. Fig. 3 shows the carbon footprint and carbon benefit for the peak period. The carbon benefit for the peak season is approximately 56,017 tons of CO_2. On the other hand, Table 3 is a comparison of the carbon benefit with popular emission activities and graphed as Fig. 3.

The carbon benefit for the peak season is almost equal to scenarios A, B, and C. For Scenario A, it involves a flight from Toronto to London. The

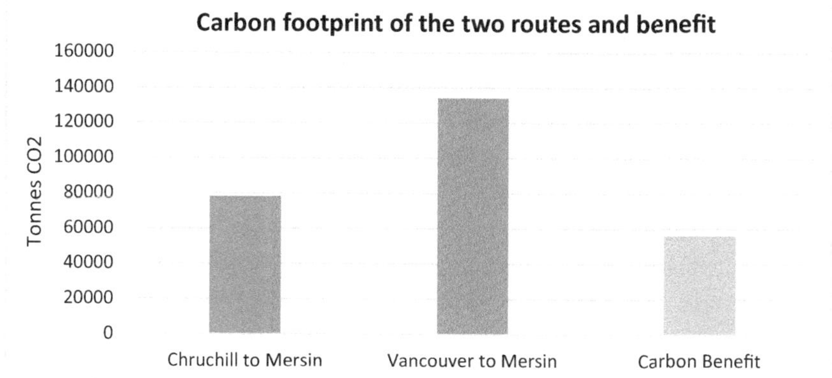

Fig. 3 The carbon footprint and carbon benefit for the peak season.

Table 3 Comparison of carbon benefit of Churchill to Mersin with common activities.

Carbon benefit from using Churchill to Mersin	Flight (Scenario A)[a]	Train (Scenario B)[b]	Car (Scenario C)[c]
56,017	56040[d]	56010[e]	55998[f]

[a]Carbon emission from Toronto to London for 500 people flying economy class return trips of 466 times.
[b]Carbon emission for train from Toronto to Vancouver for 500 people with approximately 5601 trips.
[c]Carbon emission for driving a Toyota Corolla 2019 SCV-7-2WD car from Toronto to Vancouver for 500 people making about 115 trips.
[d]https://calculator.carbonfootprint.com/calculator.aspx?tab=3.
[e]https://calculator.carbonfootprint.com/calculator.aspx?tab=6.
[f]https://calculator.carbonfootprint.com/calculator.aspx?tab=4.

assumption going into this calculation is that average an individual flying via economy class emits 0.24 tons of CO_2 for the Toronto to London route on a return flight which implies that, our carbon benefit is equivalent to a plane carrying 500 people and making approximately 466 round trips. On the other hand, Scenario B involves the use of a train. Considering the use of a train the average emission is 0.02 tons of CO_2. In terms of the carbon benefit, it will be equivalent to 500 people making trips with the train with approximately 5601 trips. For Scenario C, it involves driving a car from Toronto to Vancouver. Using, for example, Toyota Corolla 2019 SCV-7-2WD, the average carbon emission is 0.97 tons of CO_2 which implies it will be equivalent to making 115 trips for 502 different people using a similar car (Fig. 4).

The comparisons show rather interesting results, for example, Churchill-Mersin route is more environmentally friendly in terms of its carbon emission measured in terms of CO_2 emission in tons. Further, the peak season's carbon benefit was also computed and subsequently compared to major activities in the transportation industry. While the analysis is not undermining the amount of carbon dioxide released, it is, however, one of the best ways to partly address the climate change problem when combined with alternate sources of fuel discussed earlier. Similarly using the energy-based carbon emission approach (CO_2 emissions = fuel consumption × fuel emission conversion factor [Tonnes CO − emissions = liters × kg CO_2 per liter fuel/1.000]) as opposed to the activity-based carbon emission methodology, the results are similar in terms of which route is environmentally friendly. See the details of the calculations are given in Tables 4 and 5.

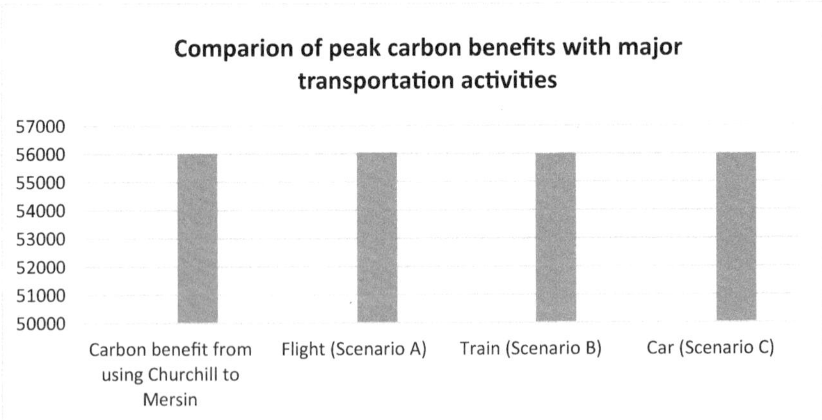

Fig. 4 Comparison of peak carbon emission benefit and major transportation activities.

Table 4 The speed and daily consumptions used.

	Supermax	
	Laden	
	Speed	Daily consumption
Full speed	13.5	31.5
Economic speed	12.0	25.0

The fuel conversion factor used is 2.9 kg CO_2/liter (Source: Anon. (2013). Methodology for calculation and declaration of energy consumptions and GHG emissions in transport services (Freight and Passengers). CEN/TC 320/WG 10. Berlin. pp. 1–66).

Table 5 The distance and total consumption used for the two routes.

Speed	Begin	End	Distance (nm)	Duration (days)	Consumption (ton)
Full	Churchill	Mersin	7176	23	724.5
	Vancouver	Mersin	12,320	39.5	1244.25
Eco	Churchill	Mersin	7176	24.9	622.6
	Vancouver	Mersin	12,320	42.8	1070

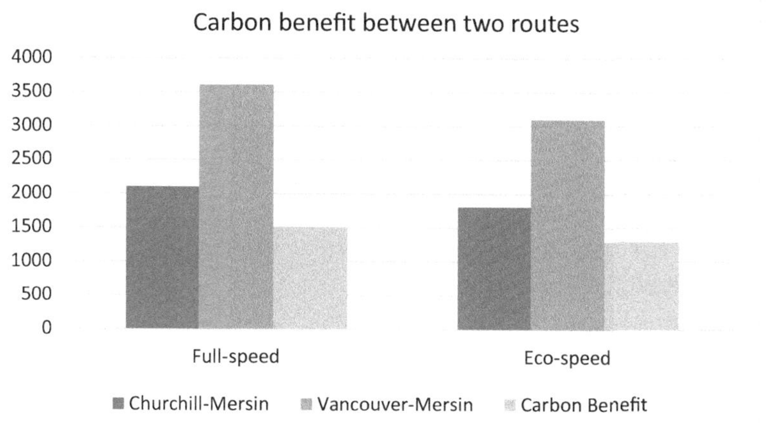

Fig. 5 The carbon footprints of the two routes and the associated carbon benefit.

The details of the two methodologies can be checked at European Chemical Transport Association (ECTA) (2011) and McKinnon and Piecyk (2010). The results of the calculations as presented in Fig. 5 show that the full-speed Churchill to Mersin route has a carbon footprint of 1751.6 tons CO_2 while the full-speed Vancouver-Mersin route produces 3014.55 tons CO_2.

This translates into a carbon benefit of 1507.28 in favor of using the Churchill-Mersin route to transport grains. Similarly, eco-speed Churchill-Mersin route saves a carbon footprint of 1297.46 tons.

3 Assessing the social impacts

One of the main social impacts identified includes demographic and geographic shifts through migration and displacement as well as urbanization and exodus from rural areas (Stephen, 2018). Another social impact is the consequences of Arctic shipping on the lifestyles which can be seen through the extensive pressure on not just the physical infrastructure but also the social ones as well. Other impacts though more health-related are analogous to social implications as well. They include hunting conditions becoming increasingly unsafe, nonavailability of indigenous food and not having safe and secure food and water. There is also a greater exposure to pollutants thereby increasing the odds of getting cancer-related diseases. It should be noted that the impact on health varies, and there have been studies showing that the effect on the indigenous communities is different to those who do not have indigenous heritage sadly in a rather negative way (Afenyo et al., 2021; Stephen, 2018).

It should be noted that these days there is a term referred to as a social license to operate which basically means that there needs to be an agreement between the companies operating in the Arctic area as well as the indigenous community. This is another indicator that would be discussed later in this section.

We cannot describe the social impact of the Arctic activities without briefly addressing the indigenous community. Sometimes it is extremely difficult to define who forms part of this group because there are several groups who inhabit the Arctic and often have different leaders. However, almost all have a common push which is the need to respect their culture and spirituality.

Climate change in the Arctic can be observed through extreme weather events, permafrost thaw which may in some cases result in ice roads that become very dangerous and unreliable, as well as severe damage to some houses, railroads (e.g., Churchill), seaports, energy as well as water supply. These happenings in most cases result in giving up the cultural and lifestyles that the local communities are used to and have been living with for many years. Activities such as hunting may be affected by the time of the ice melt, and such may also affect the breeding rate and time of animals like polar bears, walruses, and seals.

These days literature and organizations have developed a measure called the Social Return on Investment (SRI). It helps companies know how to improve their impact on society and therefore improve communication between these organizations and where they operate. The general equation used to describe SRI is

$$SRI = \frac{X - Y}{Y \text{ x } 100\%}$$

where X is the social impact value and Y is the initial investment amount.

The main problem, however, with such a measure is that when you assign a dollar value to the social impact it becomes a little problematic. To do this, the Analytical Hierarchical Process (AHP) is sometimes used, and research is still ongoing to better understand how this can be done in a fair way. For example, how will you be able to assign a dollar value to worship. A social-economic impact model has been developed by Afenyo et al. (2021) and is shown in Fig. 1. In this model, the gray nodes which include the effect of oil spills on water supply, social disruption, psychological effect, level of thrust, culture disruption, hunting disruption, and disruption of association of individuals in a community affected by oil spills are those related to social impact. The authors used a reverse estimation technique to estimate their values. For example, the psychological effect of oil spills on the populace could be quantified by estimating how much it costs to get treated by a professional psychologist when faced with the fear of losing job, marriage, family, etc., due to migration driven by an oil spill or other related incidents. The rates used are specific to the region the model is to be applied to.

In addition, Larsen et al. (2010) also conducted a study on the Arctic and came up with the following indicators for assessing how well things are going in the Arctic. The indicators include *Health and population well-being, material well-being, educational well-being, Cultural Well-being, fate control, contact with nature indicator.* In terms of health and population well-being indicators, the most used is infant mortality. This is to measure how many children in the community die before they are even born. If this number is very high in the Arctic community, it means that very likely the health of the people living there have serious challenges. One of the many challenges is encouraging healthcare professionals to go and work in these areas where some Arctic countries are very remote. An alternative solution would be to encourage indigenous people to go into the medical professions like Medicine, Nursing, and Pharmacy.

In terms of material well-being indicators, per capita *household income* was chosen as the measure. This is basically the total income of a particular household divided by the number of people in that household. The challenge however with such a measure is that it does not address the informal sector. The Arctic has a lot of its populace engaged in informal work activities. Educational well-being is assessed based on the following indicators: The proportion of students in school that are pursuing postsecondary education. In Canada, for example, there are not a lot of schools in the Arctic region and only one or two universities. This makes it difficult to train the required manpower necessary for sustaining the region. Another indicator of educational well-being is the ratio of students who successfully completed their postsecondary education to the total number of students in school. Finally, the students who remain in the community 10 years after graduation are another indicator of educational well-being. Due to the lack of opportunities and jobs, some of the people trained in these regions are likely to move to other industrialized regions, except they find a well-paying job in this region.

Cultural well-being is mostly linked to the transfer of indigenous knowledge and cultural practices as well as the ability to practice their way of worship and living freely. Indigenous knowledge is handed down from generation to generation and is often guarded and preserved for future generation. If a particular community in the Arctic is not free to live its traditional way of life, it is an indication that there are major challenges in this community and most likely coming from outside influence or activities. Arctic shipping will facilitate tourism, implying that a lot of people will have the opportunity to interact with the local people as they visit these regions which in times past were impossible to visit.

In terms of *contact with nature*, the ease of accessibility to traditional food is a major indicator of this. In addition, the consumption and harvesting of the same is another very important factor in assessing contact with nature as an indicator. Finally, the fate control indicator is a function of the fraction of indigenous people in decision-making positions. This is very essential because nobody knows the indigenous community better than the people living within it and knows what is most needed in these communities. Another contributor to the *fate control indicator* is the percentage of land and other resources that have been legally acquired in the community. For example, land for mining gold and other precious minerals needs to be properly acquired. Is the central government the only one going to get some royalties or the indigenous communities through the chiefs and elders and the community leaders. The number of individuals who speak their mother tongue in the community is another factor that helps define the fate and control indicator. Finally, one critical indicator is the percentage of expenses for public expenditure with sources from that jurisdiction (Larsen et al. (2010).

The authors of the studies above emphasized that these indicators must have the following characteristics, namely (1) data availability, (2) ease of measurement, (3) robustness, (4) scalability, and (5) inclusiveness. These and many other studies have attempted to estimate the social impact of Arctic activities. However, there is still a long way to go get accurate estimates of the social implications of Arctic activities.

4 Discussions

The changing climate and the melting of ice that comes with it present numerous opportunities both economically, socially, and environmentally. However, in literature, the social impact is difficult to assess while the environmental impact may be very visible but not easy to measure as well. In the context of sustainability, it is extremely important that the resources available today are used wisely and some preserved for future generations. These resources are not just tangible but could be intangible as well, for example, the culture of the indigenous people and the food they eat. As discussed earlier, these are captured by researchers in the form of indicators and are often difficult to measure. The Arctic and its activities can only be managed in a sustainable way if the environmental, social, and economic components are properly addressed. In the global context, these impacts need to be considered when enacting laws and formulating developmental goals for the future. Even though we cannot fully know all the impacts, the ones identified so far can form the basis for decision-making by policy-makers as well as law makers. The major challenge relating to assessing the social and environmental impacts in the Arctic continues to be limited observational data as it is sometimes extremely difficult and expensive. The study of the Arctic and its impact needs to be done in a global context and not in isolation. The Arctic is not the only region experiencing climate change and is not the only region with ice. Antarctica is another region, and we may benefit in one way or the other in terms of how social and environmental impact is analyzed in such a region. Of course, working in the Arctic is a little less challenging compared to Antarctica. However, that is not to say there are not some data that can be shared, and trends inferred.

5 Conclusion

This chapter presented the problems related to assessing the social and environmental impact of Arctic shipping and related activities. The chapter identified the difficulties we face when trying to assign quantities to social factors and suggests that more research needs to be done to have an improved way of doing this. The chapter also presents work done by Afenyo et al. (2021) and others who attempted to do this. In addition, assessing the social and environmental impact of Artic shipping and related activities needs to be done in a global context and should bear in mind the push for sustainability. The discussion of the indigenous people was the basis for another study which was conducted with the sole purpose of using different indicators to describe the social impacts. The lack of observational data remains a major challenge as well as the risk associated with measuring quantities in an evolving terrain like the Arctic. Governments and other law-makers in the Arctic need to bear in mind the social impacts when making laws and critical decisions. It is also very important the leaders of the Arctic communities are involved in every stage of decision-making concerning their community and region. Royalties from mining and other activities need to be shared in an equitable manner so the local people can fully benefit from the resources on their land.

References

Afenyo, M., Ng, A. K. Y., & Jiang, C. (2019). Climate change and Arctic shipping: A method for assessing the impacts of oil spills in the Arctic. *Transportation Research Part D*, *77*, 476–490.

Afenyo, M., Ng, A. K. Y., & Jiang, C. (2021). A multi-period model for assessing the socio-economic impact of oil spills in the Arctic. *Risk Analysis: An international Journal*, *42*(3), 614–633.

Afenyo, M., Ng, A. K. Y., & Jiang, C. (2023). A Bayesian-Loss Function (BLF) Hybrid model for assessing the Marine Liability Regime for Ship-Source Oil Spills in the Arctic. *Risk Analysis: An International Journal*, *43*(8), 1557–1571.

Atilhan, S., Park, S., El-Halwagi, M. E., Atilhan, M., Moore, M., & Nielsen, R. B. (2021). Green hydrogen as an alternative fuel for the shipping industry. *Current Opinion in Chemical Engineering*, *31*, 100668. https://doi.org/10.1016/j.coche.2020.100668.

European Chemical Transport Association (ECTA). (2011). *Guidelines for measuring and managing CO_2 emission from freight transport operations*.

IMO. (2020). *Fourth IMO greenhouse gas study*. https://wwwcdn.imo.org/localresources/en/OurWork/Environment/Documents/Fourth%20IMO%20GHG%20Study%202020%20-%20Full%20report%20and%20annexes.pdf. (Accessed 19 August 2023).

Jacoby, M. (2022). The shipping industry looks for green fuels. *C&EN*, *100*, 8. https://cen.acs.org/environment/greenhouse-gases/shipping-industry-looks-green-fuels/100/i8. (Accessed 19 August 2023).

Larsen, J. N., Schweitzer, P., & Fondahl, G. (Eds.). (2010). *The Arctic social indicators report.* Copenhagen: Nordic Council of Ministers. http://library.arcticportal.org/712/1/Arctic_Social_Indicators_NCoM.pdf. (Accessed 24 August 2023).

McKinnon, A. C., & Piecyk, M. (2010). *Measuring and managing CO_2 emissions in European chemical transport.* CEFIC. http://www.cefic.org/Documents/IndustrySupport/Transport-andLogistics/Sustainable%20Logistics/McKinnon%20Report%20Transport%20GHG%20emissions%2024.01.11.pdf.

Stephen, K. (2018). Societal impacts of a rapidly changing Arctic. *Current Climate Change Reports*, *4*, 223–237. https://doi.org/10.1007/s40641-018-0106-1.

Svanberg, M., Ellis, J., Lundgren, J., & Landälv, I. (2018). Renewable methanol as a fuel for the shipping industry. *Renewable and Sustainable Energy Reviews*, *94*, 1217–1228. https://doi.org/10.1016/j.rser.2018.06.058.

CHAPTER TEN

Indigenous knowledge and culture: How important is it for Arctic development?

1 Introduction

The International Labor Organization's *Convention 169* defines indigenous peoples as: "peoples in independent countries who are regarded as indigenous on account of their descent from the populations which inhabited the country, or a geographical region to which the country belongs, at the time of conquest or colonization or the establishment of present state boundaries and who, irrespective of their legal status, retain some or all of their own social, economic, cultural and political institutions" (Sanders, 1999, p. 2).

Indigenous people have been living in the Arctic region for thousands of years. The native residents constitute 10% of the 4 million occupants of the Arctic. The native people are roughly divided into more than 40 tribes living in all Arctic countries except Iceland. These tribes are Sami in Finland, Sweden, Norway, and Northwest Russia; Nenets, Khanty, Evenk, Chukchi, Eskimosy (Inuit/Yuit), Mansi, Selkup, Dolgan, Yukaghir, Aleut, Itelmen, and several others in Russia; Aleut, Yupik, and Inuit (Iñupiat), Aleut, Tlingit, Ahtna, Deg Hit'an (Ingalik), Dena'ina (Tanaina), Doogh Hit'an (Holikachuk), Dichinanek Hwt'ana (Kolchan), Eyak, Gwich'in (Kutchin), Hän, Koyukon, Tanana in Alaska; Inuit (Inuvialuit), Cree peoples, Deh Cho (Slavey), Denesuline (Chipewayn), Dunneza (Beaver), Gwich'in, Hän, Kaska, Sahtu (Hare), Tagish, Tlicho (Dogrib), Tutchone, and the Innu in Canada and Inuit (Kalaallit) in Greenland. However, the number of indigenous people in all Arctic nations differs significantly. For example, in Nunavut, Canada, 85% of the total inhabitants are native residents; similarly, in Greenland, a significant portion of the total residents are indigenous. In contrast, other areas, such as the Khanty-Mansi Autonomous Okrug of Siberia, have fewer native citizens (less than 2%) (Fondahl et al., 2015).

Safe and Sustainable Arctic Shipping Management and Development
https://doi.org/10.1016/B978-0-443-18849-7.00002-6
Copyright © 2024 Elsevier Inc.
All rights are reserved, including those for text and data mining, AI training, and similar technologies.

All the native tribes have different kinds of culture, history, and economic status. However, several factors, such as exposure to modern Western life, government strategies related to the Arctic, and greater access to the other parts of the world due to better transport connectivity, have impacted all these native groups. They share characteristics such as attachment to their motherland, unique language, culture, and living sources (reindeer herding, fishing, and hunting). However, their unique culture and way of life are under threat due to climate change and increased business and industrial activities in the Arctic region. To protect their culture and economic rights, the political representatives of the indigenous people have been raising their voices at national and global levels and have successfully gained international attention and recognition of their economic and political rights (https://www.arcticcentre.org/EN/arcticregion/Arctic-Indigenous-Peoples#:~:text=Arctic%20indigenous%20peoples%20include%20for,Inuit%20(Kalaallit)%20in%20Greenland).

For Arctic development, it is essential to integrate the knowledge and culture of indigenous people. The reasons for their inclusion are that these native people come from several diversified backgrounds, and each native tribe has its distinctive knowledge, culture, and living style, which they have developed over a very long period. Including these indigenous people in the Arctic, development strategies are necessary from the ethical and human point of view and essential for sustainable Arctic development.

2 Indigenous knowledge and culture in Arctic development

The United Nations Educational, Scientific and Cultural Organization explains indigenous knowledge as "*the understandings, skills, and philosophies developed by societies with long histories of interaction with their natural surroundings*" (https://en.unesco.org/links#:~:text=%C2%A9%20UNESCO,day%2Dto%2Dday%20life). Indigenous people deeply understand their land and resources. For instance, they know the distribution of resources, how the ecosystem works, and how the environment is closely related to their culture (Stevenson, 1996). Indigenous knowledge is a framework developed based on structured thinking and awareness applied to biological, physical, cultural, and linguistic circumstances. Indigenous people own this rich information and transfer it to the next generations through native languages. This source of information is generated through cultural customs and lifetime experiences that native people have gathered through observations, learning, and skills.

Indigenous knowledge has been developed through several generations and is still in the process of improvement (https://lcipp.unfccc.int/sites/default/files/2022-06/Arctic%20Indigenous%20Peoples.pdf).

Indigenous peoples' leading source for their survival has been their knowledge and awareness about their surroundings and environment. Thanks to their unique knowledge and know-how, they have successfully worked in small independent groups to gain benefits from the division of labor while developing strong social connections within and between their tribes (https://www.amap.no/documents/download/1084/inline). The following sections explain how indigenous people's knowledge and culture can be utilized for Arctic development (Fig. 1).

2.1 Environmental knowledge

The impact of climate change on the Arctic environment, including species and ecosystems, has attracted the attention of academics and practitioners. Although scientific research and their findings have helped understand the potential threat to the Arctic environment and possible policies to handle it, they present only some aspects of a complex situation. In this regard, indigenous knowledge reflects other angles of the Arctic environment in danger of the impact of climate change (https://www.cbd.int/doc/publications/tk-cc-arctic-en.pdf). Thus, the scientific communities have

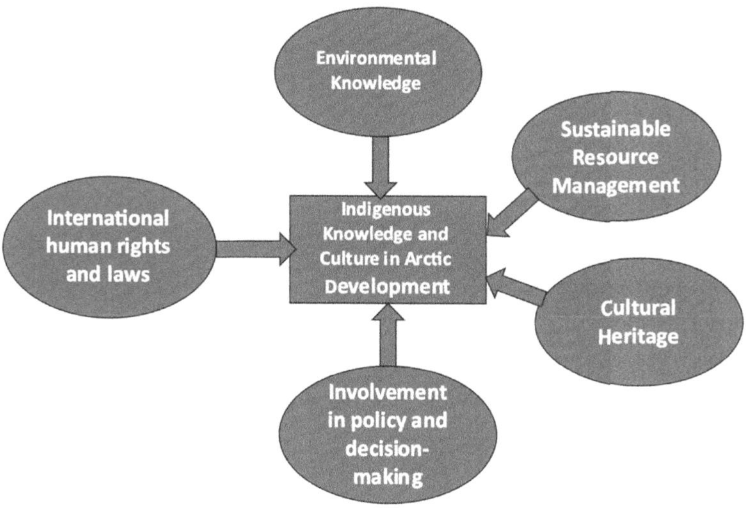

Fig. 1 Factors contributing to indigenous knowledge and culture in Arctic development.

recognized the limitations of conventional science in analyzing the complex ecological challenges and stressed the need to incorporate the environmental knowledge of indigenous people in handling and expanding the Arctic resources (Wolfe et al., 1992).

International agencies have also emphasized the need to integrate the indigenous people in essential policy implications that impact their land, culture, and ways of living. See, for instance, Agenda 21 of the United Nations Conference on Environment and Development, 1992. Moreover, several global Arctic agreements, such as the Arctic Environmental Protection Strategy (Vanderzwaag, Huebert, & Ferrara, 2001), have emphasized the role of the indigenous community's rich knowledge, which they have gathered through the generation, in implementing environmentally friendly and sustainable development policies. In this respect, the Canadian government has taken a great initiative. Its federal environmental guidelines made it compulsory to integrate indigenous knowledge into environmental impact assessment to ensure the sustainable development of the Northern area and its communities (Usher, 2000). Such initiatives are essential not only in utilizing the native people's knowledge but also in making sure that their concerns are fully acknowledged, consequently resulting in better management of environmental effects and benefits to the northern citizens. Resource extraction activities such as mining have potential economic benefits, but at the same time, these activities negatively influence the environment and ecosystem. In this respect, involving the indigenous people and their knowledge in assessing the environmental impact of resource extraction activities will benefit the policymakers and the native people (Stevenson, 1996).

Riedlinger and Berkes (2001) divided the use of indigenous knowledge into five categories that could help to understand and carry out modern science about the Arctic climate-change research:

(1) Large-scale expertise of local communities: The indigenous people, such as the Inuit in North Canada, notice the changes in climate change through biophysical variations in the level of ice in the sea, flora and fauna, glaciers, and weather. For instance, the local community of Sachs Harbor has observed the sea-ice changes in the Beaufort Sea. They used the following characteristics to explain differences in the sea ice: sea water temperature, thickness, color and phenology of ice, pressure ridge, distribution pattern, and health of animals and wind flows. The variations in the sea levels are observed by comparing these characteristics' past and present situations.

(2) Climate background: The indigenous people's knowledge and history are good sources for tracing past variations in climate, which can be used as a basis to measure climate changes. The climate history is integrated into the Inuit history of flora and fauna, travels, disasters they have faced, and harvesting of their land.

(3) Environmental research formulation: The researchers can benefit from indigenous knowledge to understand climate change from a different perspective and also in formulating the research questions. The involvement of the indigenous people in the early stages of research benefits researchers and local communities. The researchers gain by improving the scope of the study, and local communities benefit by establishing their critical role in the research process.

(4) Indigenous community's response to climate change: Indigenous knowledge explains how they have responded and adapted to the climate changes and the hurdles from economic, social, and cultural points of view to adaptations.

(5) Inspection of the environment by the indigenous community: Indigenous knowledge shows a systematic evaluation and Inspection of the environment. The scientific community and policymakers can benefit from indigenous knowledge by involving them in environmental monitoring projects.

Indigenous people are a reliable source to monitor the changes in several aspects of the Arctic biodiversity due to their traditional activities for living, such as hunting, herding, farming, and fishing. For instance, in Gambell, Alaska, the United States, some native hunters noticed the following changes in animal lifestyle:

I. Walruses, who used to live in large groups on a big part of the ice, now live individually on small portions during the Spring season.

II. Some animals have shifted their migration route from the western side (Strait) and northwest of Saint Lawrence Island to the eastern side.

III. It is challenging to fish tomcod and blue cod in winter.

IV. On the positive side, climate change also has some good impacts on animals and plants. For example, the health of marine mammals has improved while there is an increase in the population of the bowhead whale. In addition, harvesting seasons for whales and salmon have changed (fall/winter for whales and summer for salmon) and the cod can now be found in local ocean. One other positive impact is the fact that there is an increase in growth of plants and roots collected for food (https://www.cbd.int/doc/publications/tk-cc-arctic-en.pdf).

Indigenous people also use cultural techniques for sharing and transferring information about environmental changes and climate adaptation to the next generation. These cultural techniques consist of the names of particular places, highlighting their ecological situation. When the environment changes later, the site's name indicates the change in the environment and ecology and the consequent adaptations that were made in response to these changes. For instance, near Iqaluit, Nunavut, an area is named "Pissiulaaqsit," which means "the location where guillemots are not found." as a testament to this drastic change. Thus, the name indicates that guillemots existed in that particular place in the past and no longer reside there. Similarly, an area is known as "Ullivinirkallak,"meaning where walrus meat was stored in the past. Since Ullivinirkallak no longer does the same function, it may show a change in the cold temperature of the area. Another place in the Arctic, "Qimirjuag," is a vast area covered by snow almost yearly. The area is rich in producing large quantities of berries, and the berry pickers get the opportunity to analyze the change in the amount and size of the snow in the field. Furthermore, polynyas—an area of open water surrounded by ice and migration routes—are also helpful in understanding the current and future changes in the movement and level of the sea. All the abovementioned cultural techniques of indigenous people can provide a guideline for policymakers and researchers to evaluate the changes in the climate and ecosystem (https://www.amap.no/documents/download/1084/inline).

2.2 Sustainable resource management

Indigenous peoples' development concept significantly differs from what modern and advanced nations consider. In general, they highly respect nature and the efficient use of natural resources and prefer to manage the land by involving the local neighborhood (Magni, 2017). They dislike selling their natural resources and emphasize community development. They apply their traditional knowledge and customs for sustainable production, consumption, conservation, and management of their natural resources (Hiwasaki et al., 2014). Examples of sustainable economic activities carried out by indigenous people include farming on a rotation basis, switching cultivation (also known as swidden agriculture), animal husbandry with domesticated animals, fishing, and agroforestry (Magni, 2017). Indigenous people utilize diverse resources to ensure a sustainable and adaptable life. They know specific processes and methods based on centuries-old experiences that ensure environmentally protected and cost-effective economic activities. These activities are carried out in such a way that ensures the provision of food to the local community while at the same time keeping the diversity of local plants (Nakashima et al., 2012).

Indigenous people in the Arctic hunted, fished, and collected all wild natural resources sustainably for thousands of years (https://www.wilsonquarterly.com/quarterly/the-new-north/arctic-conservation-in-the-hands-of-indigenous-peoples). For instance, a fishing method known as the "Straits Salish Reef net Fishery" is acknowledged as a sustainable technique that indigenous people have used for thousands of years in Canada. The fishing method applies the experienced Reef Net technique to catch salmon by seizing their migration routes in the Salish Sea. Indigenous people (Salish communities) use their knowledge of salmon migration routes to implement this unique method. Since the fishing technique does not involve burning fossil fuel and releasing unnecessary bycatch, it is considered a sustainable approach. Another excellent example of sustainable fishing is "Mi'Kmaq Lobster Fishery," which indigenous people in the coastal region of eastern Canada (Mi'Kmaq people) apply. They manage and regulate their ocean resources based on three beliefs: one should be kind to nature, generous to share the resources, and have a spiritual relationship with nature. They have their philosophy to manage the resources known as "Netukulimk," which recommends that the consumers use no more than the required resources so that enough is left for the next generations. Finally, another sustainable fishing method is used by the Northern Pacific Rim indigenous people. They have been fishing and catching salmon for more than 10,000 years. Like the Mi'kmaq people, they follow the fishing rule that ensures sustainability by avoiding overfishing and maintaining enough population of salmon for future use (https://www.shakeuptheestab.org/post/indigenizing-our-perspective-on-fisheries#:~:text=Methods%20of%20catching%20fish%20for,vastly%20more%20sustainable%20(4).&text=First%20Nations%20and%20Inuit%20communities,thousands%20of%20years%20(5)).

However, the authorities responsible for managing and making policies related to the resources often have minimal interaction with indigenous people to utilize their knowledge and benefits for the sustainable management of the natural resources in the Arctic. Currently, indigenous people use their expertise and information locally, and government authorities rarely gain benefits from this rich skill possessed by the local communities. For instance, local reindeer herders use unique and practical methods that national policymakers can learn and apply. However, local herders complain they struggled with the complex democratic setup to transfer their knowledge. The difference between scientific knowledge and indigenous knowledge (e.g., local hunters' experience) is that the former emerges from education while the latter is obtained from building close contact with nature. Moreover, although scientific knowledge has a monopoly in natural resource management, biodiversity continuously falls, indicating a need to integrate local and indigenous knowledge with scientific knowledge. Thus, close collaboration and knowledge sharing among the local hunters, fishers, and government staff is recommended. Such a collaboration will also encourage indigenous people's rights to manage and govern their resources and land (https://norden.diva-portal.org/smash/get/diva2:791816/FULLTEXT03.pdf). Following the indigenous people's sustainable management practices will help protect the natural resources and ensure ecological and economic sustainability.

2.3 Cultural heritage

The culture of the indigenous people is a unique and essential component of the advanced civilization. The main features of their culture are that they have complete control and knowledge about the Arctic landscapes, have adjusted to extreme weather and natural conditions, have developed their values, and live in peace with nature without damaging the ecology and biodiversity. The importance of culture in Arctic development is evident because culture and cultural aspects are mentioned in the definition of sustainable development by the Arctic Council. The unique characteristics and variability of the indigenous culture also contribute to the country's economic development by attracting tourists worldwide (Lempinen, 2019). However, their distinctive culture is under threat due to globalization and industrial progress in the North. The several characteristics of the culture of the natives, such as their language, lifestyle, reindeer husbandry, hunting, and fishing, are under threat due to exposure to the modern world and industrial advancement (Zhozhikov, 2022). And despite the culture's strategic importance, Sweden is the only country out of the eight Arctic countries that has mentioned "culturally sustainable development" in its Arctic policy. Moreover, the approaches and governance structure of the Arctic countries' central government and even the international political system usually do not match the cultural values and social setup of indigenous people. Some political decisions, such as a European Union ban on importing seal products, have been regarded as a threat to Northern cultures, values, and lifestyles by Canada and Denmark (Lempinen, 2019).

Indigenous people possess a good share (they speak 4000 languages) of the world's languages compared to their percentage of the world's population (only 6%). About 40% of the worldwide languages, which constitute a high proportion of indigenous languages, are under threat and most likely will become extinct. A minor percentage of the world's languages are learned through a well-established education process and are protected and developed through digitalization. The indigenous people in the Arctic speak roughly 40–90 languages, and many of them are under threat of eradication (https://arctic-council.org/news/revitalizing-arctic-indigenous-languages/). If the indigenous language fades away, it will likely mean the elimination of the rich knowledge gathered by the native community. Indeed, the global culture and biological diversity could be at risk, too (https://www.unesco.org/en/articles/cutting-edge-indigenous-languages-gateways-worlds-cultural-diversity). For example, the Sami language uses more than 300 terms to explain snow and changes in the snow pattern. To protect their language, several local and international organizations have taken measures to revitalize their languages. For instance, to protect and advance the indigenous languages, the United Nations declared 10 years starting from 2022 as the "International Decade of Indigenous Languages," highlighting the importance of indigenous languages not only to the indigenous culture but also to the world culture and diversity (https://arctic-council.org/news/revitalizing-arctic-indigenous-languages/).

Similarly, reindeer herding is a source of earning, a way of life, a link to ancestral roots, and a means of connection with other community members. In some communities, this is an integral part of the identity of the community and this in turn help harmonize the connection between the native people and animals. It is a source of two-way interactions between the native people and the ecosystem, describing the situation where indigenous people depend on animals and vice versa. In some areas, herding is done through the traditional methods applied for centuries, such as reindeer sledges and riding the reindeer. Climate change and the use of advanced technologies (e.g., motorcycles, snowmobiles, helicopters) in reindeer herding threaten traditional herding, which might affect the culture and connections of the native people with their community and nature (https://lcipp.unfccc.int/sites/default/files/2022-06/Arctic%20Indigenous%20Peoples.pdf). Hence, promoting and protecting indigenous languages, values, and lifestyles is essential for cultural heritage and keeping a diverse and dynamic Arctic community. Their rich culture is crucial in a more inclusive and holistic approach to Arctic development.

2.4 Involvement in policymaking and decision-making

The Arctic Council (AC), founded in 1996, is the main decision body and regional organization. The AC consists of eight Arctic nations and permanent representatives of indigenous people. As permanent participants, the following organizations represent indigenous people in the AC: the Arctic Athabaskan Council; the Aleut International Association; the Gwich'in Council International; the Inuit Circumpolar Council; the Russian Association of Indigenous Peoples of Siberia and the Far North; and the Saami Council (Sharapova et al., 2022).

As discussed earlier, indigenous people are a reliable source of knowledge for Arctic development and sustainably managing the Arctic ecosystem and biodiversity. However, as explained by Coote (2016), they are also the most affected group due to climate change and its impact on the Arctic. Considering the influence of climate change on native peoples' lives, the AC emphasized the need to involve the indigenous people in its Ottawa Declaration. The declaration proposed a framework to integrate the indigenous people in various projects, including the specialized working groups responsible for a significant portion of the AC's work and its responsibilities. However, despite their representation in the governing body (i.e., AC), indigenous people do not have the authority to influence and make any decisions. The native people have been and continue to be largely excluded from key governance and management decisions (Fraley, 2021; Sharapova et al., 2022).

One example of excluding the rights and opinions of indigenous people from critical decision-making is the Ambler Road Industrial Project, which commenced in July 2020 despite strong opposition from the native people in Alaska. The road project has several potential risks for the native people, so it will go through the National Park in Alaska, thus affecting the wildlife; it could expose the marine resources (inconnu, salmon, and northern pike) of the Kobuk River to potential disasters related to oil extractions in the region. Another example of a decision taken without consulting the indigenous and ignoring their preferences is the drilling for oil extraction in the Arctic National Wildlife Refuge of Alaska. The main drawback of this drilling activity is the negative consequence on the population of the Porcupine caribou herd, which is the primary source of food for the Gwich'in and Iñupiaq groups of indigenous people (Fraley, 2021). Thus, it is essential to integrate indigenous people and their opinions into policymaking and decision-making processes to understand the local perspective better. This approach will produce more effective development policies that are more sensitive and responsive to local demands and needs.

2.5 International human rights and laws

Since the early 1980s, the rights of indigenous people have attracted the attention of international law and policymakers, mainly in response to the active movements initiated by several stakeholders, including indigenous people, civil society, and governments at the local, regional, and global levels. The United Nations human rights system, including its channels, laws, and regulations, has been the primary source for protecting the rights of indigenous along with other international forums such as the United Nations Working Group on Indigenous People. The Human Rights Council and the United Nations Permanent Forum on Indigenous Issues later supported their work (United Nations, 2013).

The United Nations Declaration on the Rights of Indigenous Peoples (UNDRIP) considers the following significantly necessary rights of indigenous people: self-reliance, self-government, the right to their native land and region, and the right to keep and follow their culture, values, and lifestyle. The rules declared by the UNDRIP are not legally authentic; however, these rules reflect some parts of international law, including the values of autonomy that are well grounded in international law. In this respect, UNDIRP's rules play a vital part in protecting the rights of indigenous people. Canada accepted and supported the rules set by UNDIRP; however, Russia is against them and has not accepted them. In addition to UNDRIP, the other nonbinding legal agreements that acknowledge the inclusion of indigenous people and their knowledge in sustainable development are the Rio Declaration on Environment and Development and Agenda 21. International human rights law and the protection of the environment are relevant, too (Sharapova et al., 2022).

Although the United Nations has given the indigenous people the right to self-determination, this right does not grant them the opportunity to claim an independent state. However, it is the responsibility of the governments to give their native people opportunities to be involved in local political decisions, which is regarded as right for "self-government of the indigenous people within their regions." Among the indigenous people in the Arctic, Sami people were the first who were given tremendous offers for political involvement, such as the foundation of the Sami Council in 1956. Later, Norway, Sweden, and Finland founded the Sami Parliaments in 1989, 1993, and 1996, respectively. Although these parliaments do not have legal authorities, they are excellent forums to express the local people's opinions and interests, which are incorporated into the national decision-making process. Also, Greenland, as a part of the Kingdom of Denmark, took a bold initiative in 2008 by conducting a referendum to give the authority of self-government to its native people, the Inuit. Their self-government has given power to the government of Greenland to manage judicial matters, policymaking, and natural resource management. Since 80% of the population of Greenland consists of Inuit, this self-government decision is a big step toward becoming the first sovereign Inuit-governed nation (Byers, 2013).

In the case of the United States, the rights of the indigenous were recognized by the government by adopting the Alaska Native Claims Settlement in 1971. This settlement has provided more safety and security for the oil activities and some autonomy to the indigenous people by allowing self-government to some extent. The settlement allotted $1 billion and founded 12 regional for-profit organizations that were given rights for resource extractions. These initiatives have created significant continuous revenue for the region, especially for about 11,000 Inupiat, who were supported by the Arctic Slope Regional Corporation (ASRC). In Canada, native Inuit negotiated several Land Claims Agreements, giving them some self-government authority. The Land Claims Agreements include the Inuvialuit Settlement Region in the Yukon and Northwest Territories to Nunatsiavut in Labrador. In contrast, in Russia, as argued by Byers (2013), the political rights of indigenous people have not yet been truly recognized and have been largely excluded from political participation.

3 Conclusion

This chapter discusses the need to include the knowledge and culture of indigenous people in the Arctic development. Since indigenous people have lived in the Arctic region for thousands of years, they have rich knowledge and culture that are excellent sources for policymakers to formulate sustainable development approaches for the Arctic and complement the modern and advanced scientific research and knowledge that are insufficient to articulate regional development plans. On the other hand, indigenous people are the ones who are affected by climate change in the Arctic region, which also makes it necessary to incorporate their knowledge and culture for an adaptable and resilient Arctic region.

The five factors that make it necessary to include native people's knowledge and culture in the Arctic development are indigenous knowledge of the Arctic environment, their ability to manage the natural resources sustainably, their cultural heritage, their involvement in policymaking and decision-making, and their recognition of their rights in international laws. Indigenous people have been involved in traditional and unique activities to earn a source of living, such as hunting, herding, and fishing. These activities have developed the ability to understand and monitor nature and changes in their environment and surroundings. The native language also contains information that can be used to identify the changes in the environment and ecology and the consequent adaptations that were made in response to these changes.

Indigenous people consider development from a different perspective than what modern and advanced nations do. They greatly respect nature and have used sustainable techniques to hunt, fish, and collect natural resources. Preserving and following the indigenous people's sustainable management practices are necessary to save natural resources and ensure ecological and economic sustainability. To protect indigenous knowledge, it is critically important to safeguard their culture, too. The unique attributes of their culture are that they have complete access and information about their native land, know how to adapt to severe weather and natural surroundings, develop their customs, and live in harmony without destroying the rich ecology and biodiversity of the Arctic. All these aspects of their culture could significantly contribute to Arctic management.

In addition to utilizing their knowledge and culture, it is equally important to incorporate indigenous people and their opinions into policymaking and decision-making processes to fulfill their needs and requirements. This strategy will generate more effective Arctic development policies. Indigenous peoples' rights are expressed in several global agreements and declarations. Thus, for sustainable Arctic development, the Arctic nations are responsible for involving indigenous people in development processes; otherwise, it can violate international human and environmental protection rights. All Arctic states, except Russia, have taken different measures to involve the indigenous people in political participation and representation. Last but not least, indigenous knowledge and culture are integral to ensuring sustainable, inclusive, and responsible progress in the Arctic. Any regional development policies should consider indigenous communities' unique opinions, experiences, and aspirations to ensure a balance between traditional ways of life and modern advancements.

References

Byers, M. (2013). *International law and the Arctic. Vol. 103*. Cambridge University Press.

Coote, M. L. (2016). *Environmental decision-making in the Arctic council: What is the role of indigenous peoples?* (Doctoral dissertation).

Fondahl, G., Filippova, V., & Mack, L. (2015). Indigenous peoples in the new Arctic. *The New Arctic*, 7–22.

Fraley, K. (2021). *Engaging Arctic indigenous communities meaningfully in decision-making affecting their food security*. East-West Center.

Hiwasaki, L., Luna, E., & Shaw, R. (2014). Process for integrating local and indigenous knowledge with science for hydro-meteorological disaster risk reduction and climate change adaptation in coastal and small island communities. *International Journal of Disaster Risk Reduction, 10*, 15–27.

Lempinen, H. (2019). Indigenous cultures, local lifestyles? 'culture' in the northern strategies of the eight Arctic states. *Research in Arts and Education, 2019*(1), 13–24.

Magni, G. (2017). Indigenous knowledge and implications for the sustainable development agenda. *European Journal of Education, 52*(4), 437–447.

Nakashima, D. J., Galloway McLean, K., Thulstrup, H. D., Ramos Castillo, A., & Rubis, J. T. (2012). *Weathering uncertainty: Traditional knowledge for climate change assessment and adaptation*. Paris/Darwin: UNESCO/UNU.

Riedlinger, D., & Berkes, F. (2001). Contributions of traditional knowledge to understanding climate change in the Canadian Arctic. *Polar Record, 37*(203), 315–328.

Sanders, D. E. (1999). Indigenous peoples: Issues of definition. *International Journal of Cultural Property, 8*(1), 4–13.

Sharapova, A., Seck, S. L., MacLeod, S. L., & Koubrak, O. (2022). Indigenous rights and interests in a changing Arctic Ocean. *Arctic Review on Law and Politics, 13*, 286–311.

Stevenson, M. G. (1996). Indigenous knowledge in environmental assessment. *Arctic*, 278–291.

United Nations. (2013). *Indigenous peoples and the United Nations human rights system, Fact sheet no. 9/rev.2*. New York and Geneva: United Nations.

Usher, P. J. (2000). Traditional ecological knowledge in environmental assessment and management. *Arctic*, 183–193.

Vanderzwaag, D., Huebert, R., & Ferrara, S. (2001). Chapter 12 The Arctic Environmental Protection Strategy, Arctic Council and Multilateral Environmental Initiatives: Tinkering while the Arctic Marine Environment Totters. *The law of the sea and polar maritime delimitation and jurisdiction*. Leiden, The Netherlands: Brill | Nijhoff. https://doi.org/10.1163/9789004482029_022.

Wolfe, J., Bechard, C., Cizek, P., & Cole, D. (1992). *Indigenous and Western knowledge and resource management systems*. Guelph, Ontario: University School of Rural Planning and Development, University of Guelph.

Zhozhikov, A. V. (2022). Digitalization of the cultural heritage of the indigenous peoples of the Arctic. In *European proceedings of social and behavioural sciences*.

CHAPTER ELEVEN

Perspective of frontline individuals involved in Arctic works and research

1 Introduction

The Arctic possesses a beautiful and serene environment. The fauna and flora in this region is like no other around the world. To transport people in Russia to the North Pole, *50 years of victory*, a Russian Icebreaker makes this possible by moving approximately 128 crew members virtually every year to the destination. It is a pride and a time to enjoy the awesome Arctic region. These crew members transported to the North Pole have the opportunity to experience a different type of environment from what they are normally used to (Brigham, 2000; Frolov, 2015).

The Canadian Arctic Watch Wilderness Lodge located in Yellowknife, NT, Canada, is an amazing site to see the beauty of the Arctic from a new level. This is an amazing tourist attraction site, which needs to be further developed to attract more visitors. In addition, there are many other tourist attractions in the area that require modernization and development (Dawson et al., 2007; Huddart & Stott, 2020). In this case, the Ilulissat Icefjord, which is a well-known UNESCO Heritage site, is an ideal place to see one of the world's oldest active glaciers. This is a perfect place to see whales and seals (Schiøtt et al., 2022; Schumann et al., 2012). If we love the sound and beauty of birds, then there is a magnificent place on Coburg Island in Canada called Cambridge Point which is probably the place to explore in the Arctic (Gaston & Bradstreet, 1993; Høye & Sikes, 2013). Spitsbergen is the Arctic capital for wildlife, and we are likely to see polar bears, walrus, and reindeer (Jakubas et al., 2017; Wråkberg, 2006). All these places are becoming more accessible because of climate change, causing shipping to such places to become a little easier (Afenyo et al., 2021).

During the Russo-Ukrainian conflict, the other Arctic countries initially believed it would not affect Arctic cooperation. However, as the conflict

Safe and Sustainable Arctic Shipping Management and Development
https://doi.org/10.1016/B978-0-443-18849-7.00015-4
Copyright © 2024 Elsevier Inc.
All rights are reserved, including those for text and data mining, AI training, and similar technologies.

persisted, Russia was suspended from the Arctic Council. The perspective of the professionals during a webinar is that it is a "temporary pause" of the Russian membership (Wilson Institute (WI), 2022). Inevitably, this would affect the potential cooperation in dealing with any consequences from oil spills in international water and rescue operations in case there is any accident in the Arctic. Nonetheless, it becomes obvious that this goes far beyond governance and cooperation, but rather national security issues. The Arctic therefore might become a battlefield for supremacy in the forthcoming years, and whoever possesses the best technology would likely have an upper hand in deciding the region's future.

While experts are discussing the impacts of the war, they reiterated that if the Arctic is to be a global waterway for container shipping, the war has worsened things. Energy cargos are normally heading to the East (Wilson Institute, 2022). The major shipping companies that have closed businesses in Russia have affected Russia and, of course, global supply chains (Wilson Institute, 2022). Businesses are likely to be affected, especially in China, which views the Arctic as a major energy, shipping, and business hub. The conflict therefore might change that for a while. Probably, countries raising their hope on the use of Arctic shipping may need to adjust their expectations a bit until the conflict is resolved.

The president of Kiribati, Anote Tong, stated clearly in his address to the UN where he shared his experience from visiting the Arctic for himself and concluded that there is more harm being done to the Arctic and requires urgent protection of the Arctic. Similarly, Micronesia, an island state in the Pacific Ocean, happens to be the third highest registry of offshore oil rigs and ships. Even the famous Arctic oil rigs by Shell, the Polar Pioneer and Transocean Spitsbergen (operated by Statoil), were registered there. This did not in any way deter the foreign minister from moving forward to ban further registration of oil rigs because of the impact of climate change on the survival of the Island (Stephen & Knecht, 2017).

From the discussions, it is very clear that depending on the interests of individuals, institutions, and countries, they perceive the Arctic from a different perspective. Interests come first, and anything follows after. The next sections will describe the views of people on the Amundsen expedition and others attending a workshop on climate change and the Arctic.

2 The *Amundsen* expedition

The author of this book was on board the Amundsen, the Canadian Icebreaker, during an Arctic expedition. He took the opportunity to seek the perspective of some of the crew in the Arctic. The crew ranged from scientists, ship captains, engineers, students, the Coast Guard, and an indigenous person as well. Details of some of the data collected have been used in the publication of Panahi et al. (2023) and Afenyo et al. (2021). For the ship captains, their perception has more to do with the challenges of navigating the Arctic. They made the need for an updated system of navigation very clear. The current instruments are a bit outdated. This will however change with new icebreakers coming onboard soon. In addition, they are worried about the patrol of foreign countries of Canadian waters. Despite the challenges, they have managed many voyages every year through the Northwest Passage through the Amundsen Science Project. This project started around 2002 and was sponsored by the Canada Foundation for Innovation as well as other partners. To the engineers, the expeditions have opened them up to understanding different weather implications; however, the day-to-day work on the ship continues to be the same, and not much has changed. The scientists on the other hand see an amazing opportunity to see firsthand what the ecosystem of the Arctic looked like by collecting and analyzing data from this region to understand some of the behaviors of the organisms in the Arctic region. Another government scientist continues to do mapping of this region to understand what is underneath the surface and what could potentially be some of the risks and prospects for natural resource exploration. Researchers continue to observe the rate of ice melting almost every year when the expedition takes place. According to the crew, they have observed over the years how the ice continues to melt and formerly ice-covered areas are becoming ice-free.

The indigenous people being interviewed, in general, are of the view that the Arctic is unique and should be left alone. In addition, even though the effect of climate change would be positive in terms of transportation by the increasing supply of essentials to this region, allowing healthcare professionals and other professionals to live and work there easily and the environment and cultural aspects need to be sustained and kept pristine.

3 The CCAPTIA conference

Through the first International Workshop on Climate Change Adaptation Planning for Ports, Transportation Infrastructures, and the Arctic (CCAPPTIA) workshop held on May 3–4, 2018, in Winnipeg, Manitoba (MB), Canada (www.ccapptia.com/18conference), the views of researchers and practitioners on the Arctic and climate change were compiled. This was done through a workshop-type symposium. Many of the points being summarized below have been further discussed in the second CCAPPTIA conference held on June 28–30, 2021 (www.ccapptia.com/21conference, held online due to the COVID-19 pandemic). However, in the following sections, we focus on the major discussion points during the first CCAPP-TIA conference due to its pioneer nature.

3.1 Climate change perspectives

The presentations mainly focused on issues related to climate change adaptation to ports and Arctic shipping as well as oil spills. Presenters came from different backgrounds. A representative from the United Nations Conference on Trade and Development (UNCTAD) explored how climate change impacts and adapts to the key coastal transport infrastructure (ports and airports) in the Caribbean. The impact of climate change is very visible in this area of the world, and unless something is done soon, we might expect more floods and other natural disasters to affect the infrastructure in places like the Caribbean. According to a researcher from the University of Manitoba, port decision-makers are aware of potential climate change impacts and feel that more adaptation actions should be undertaken. He said they were, however, skeptical about their effectiveness and value. His presentation also showed the complexity of the problem of climate change in terms of planning the involvement of port decision-makers under the current circumstances.

A study conducted by a researcher from the University of Rhode Island (RI) explored the Port of Providence (RI) stakeholder community's perceptions around the responsibility to lead in resilience planning and systemic transformational change, specifically evaluating how differences in perceptions between stakeholders and assumed leaders may influence adaptation. His presentation highlighted the need for pre-planning dialog to develop a consensus and build momentum for resilience investment strategies.

Another researcher from the University of Manitoba presented her perspective regarding the use of virtual reality (VR) and how it would affect us in understanding climate change, including Arctic shipping and development. She addressed the effects of VR on four theorized dimensions of psychological distance—temporal, social, geographical distance, and uncertainty—toward climate change. According to her, lower psychological distance would lead to higher levels of concern. In addition, she argued that VR would lower psychological distance, while the level of concern would be positively related to consumer's preparedness to act on climate change. The Mayor of a town in northern Canada gave his perspective on the effect of climate change on Arctic townships and ports. He stressed the need for massive investments, especially as the transport infrastructures were often prone to erosion due to the extreme conditions in the region. He used his town as an example, highlighting that his town was a major site for tourism but often seriously affected by vulnerable transport infrastructures, thus requesting some injection of money from regional, federal, and private agencies.

One of the attendants from the private sector was of the opinion that since we already had a well-established engineering approach to climate change, there was a need for a supply chain management (SCM) approach. There was also the question of who would take the mantle of leadership for climate change-related issues and what the incentive for this would be. Furthermore, it is also important to have a scale in mind. Thus, whether we are going to be dealing with a regional issue or is going to be global in nature, finally, he suggested that there is an urgent need for an integrated approach. Thus, it would require that social scientists, politicians, economists, and engineers work together.

A researcher on climate change from the UK also presented his perspective, and that is the need for models developed to be guided by policy for the value to be appreciated. In this regard, he stressed the need to build a connection between natural scientists and social scientists. For model creation, he advised that there is a need to create a framework first and then fill in the data. Data may be engineering, economic, and other aspects in nature. This researcher also presented an interesting opinion on different disciplines' approaches to climate change. He clarified that different disciplines view problems through different lenses, and therefore their corresponding solutions differ, for example, environmentalists versus engineers and engineers versus social scientists. Because of the nature of climate change problems, there is a need to learn from other fields and jurisdictions. Finally, he suggested that an optimization perspective would be cost rationalization and the incorporation of the positive impact of climate change in models and policies.

A Canadian researcher on the subject matter suggested that it is essential climate change could be measured through a meaningful index. He, however, admitted that it would be complicated. He is also of the view that it is imperative to re-educate people about the climate change. Currently, the approach adopted is to scare people; thereby, working on their psyche would help a lot. The climate change effect in the Arctic could be a great starting point, as there are some positives here. If people get educated on the implications of climate change, they can better contribute to it.

3.2 Arctic perspective

During the CCAPPTIA conference, one of the world's renowned Arctic researchers, Dr. David Barber from the University of Manitoba, stated that his perspective of the Arctic is one of many opportunities, especially as the ice continues to melt. He backed this up with over 10 years of data, showing how much the ice has reduced in thickness. However, a researcher from Newfoundland in Canada presented a picture of the numerous risks in the Arctic, especially when it pertains to operations. He is worried about how effectively we can evaluate these risks in a timely manner. Even though he proposed some solutions, he reiterated that there is a lot that needs to be done before we can reasonably address the risks associated with operating in the Arctic especially when it has to do with spills. Like the concerns of the researcher from Newfoundland in Canada, two other researchers from the University of Manitoba were worried about the potential risks of oil spills. However, these risks are not only operational but also environmental, social, and economic. The worrying questions circulated about whether our risk control options were cost-effective in the Arctic. In addition, there is a concern about how we would assess the socioeconomic impact of an oil spill in an environment like the Arctic. Their perspective is that it is difficult and requires more data and sophisticated tools and methods.

A Chinese-based academic has an interesting perspective on how China, a near Arctic country, views the Arctic. He describes this through the Impacts of China's Arctic Policy (CAP) on the East Sea Economic Rim. Furthermore, he gave an overview of the CAP and its possible impacts on the East Sea Economic Rim (ESER) region including China, Russia, North and South Korea, and Japan. His presentation has provided useful insights into a better understanding of the possible changes in port connectivity and economic and trade corridors across the associated regions due to the introduction of the CAP. In this case, China views the Arctic as very critical to trade and for that matter GDP growth. Due to how critical the Arctic is to China's growth, another Chinese researcher is of the opinion that there is a lot of investment in Arctic research, especially the safety of Arctic shipping and the world would benefit a lot from receiving some of these with them. In the same light, both the Chinese researchers agreed that ship trajectory analysis is very important; in doing so, data recorded in 2016 in the north latitude 66.67° in the Arctic, with the passage through the Pari Channel, the Dove joint Straits, and the Vilkitskogo Strait, were analyzed. They highlighted the essence of this analysis as (1) to identify the types and quantities of vessels going through the Northwest Passage and the Northeast Passage and (2) it is found that one passenger vessel went through the Pari channel, and the other ships passed through the Dove joint Straits. The passenger vessels were mainly assembled in the Davis Strait and Baffin Bay area. The navigable periods of the Northeast Route are May, July, August, September, October, November, and December, among which the number of vessels during August, September, and October is the most, according to his presentation.

Finally, a researcher presented his perspective on pollution due to black carbon in the Arctic. According to this researcher, increased black carbon (BC) emissions from fuel oil combustion from shipping could have a noticeable local warming effect—warming from the dual effects of BC solar radiation absorption and albedo reduction from deposition on snow and ice. However, he noted that many uncertainties remain on the actual impacts and questioned the kinds of policies that would best address these impacts.

4 Discussions

Based on the above discussions, Fig. 1 captures the problems in the form of the drivers of climate change as well as the vulnerabilities. These vulnerable areas include coastal transport, inland transport, and hinterland connectivity. The latter is a serious problem, especially for those in the remote areas of the Arctic where it is extremely expensive to provide high-speed access. The areas of vulnerability include air, sea, road, and rail infrastructure and the cruise industry. Such vulnerability was exemplified in 2017–18 with the disruption of the (only) rail tracks connecting the town and port of Churchill, MB (the largest port in northern Canada) with southern Canada, and the economic and social hardship was such a disruption has brought to the region and its communities. Indeed, the solutions to these problems are governance, business, and engineering related.

Fig. 1 captures some of the strategies, mainly in the form of risk management strategies, adaptation and mitigation, contingency planning, and resilience, among others. These eventually lead to potential research areas and problems that we must address carefully. From the perspective of professionals, climate change and the Arctic are interrelated. Solving problems is related to one, and leaving the other is not necessarily the best way to go. Some potential solutions to these problems would be to encourage

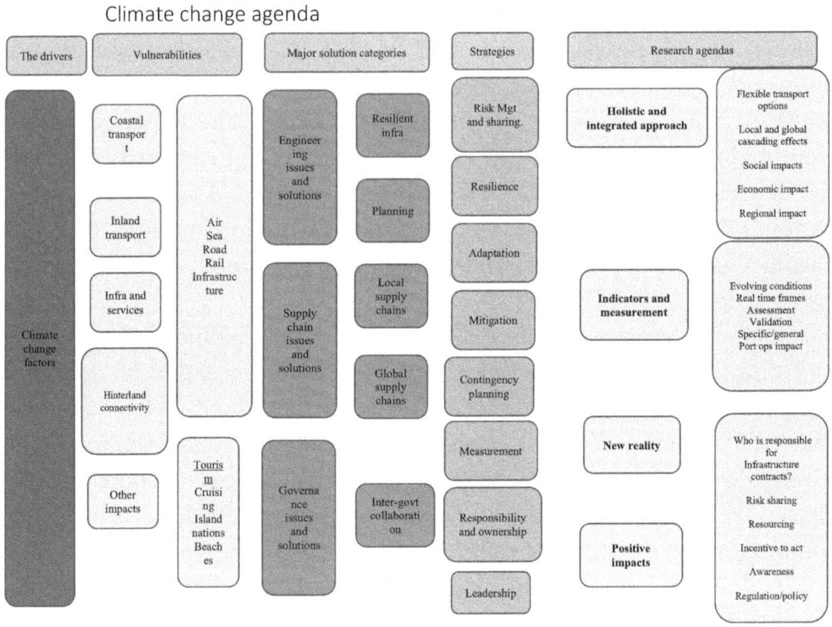

Fig. 1 Climate change and the Arctic problem and a possible solution from the perspective of professionals.

collaborations. In doing so, it is important that groups working on such problems make it clear what data are required to address the critical problems related to these happenings.

Oil spills remain a major concern from the perspective of these people, especially when it comes to pollution, and it is important that relevant groups take into consideration the policy and regulations for the application of dispersants and in-situ burning. Furthermore, there is a need to track industrial trends in the usage of different types of oil and take into consideration the possibility of the spill of different products. The harmonization of policy regarding oil spill research in the Arctic would go a long way to address the bigger problem of the recovery and restoration of affected regions. Solving any of these problems would require a multidisciplinary approach, and in doing so, there is a need for thorough feedback from colleagues from different sectors, thereby creating an effective communication strategy among all participants. In this regard, it is important to note that governmental issues would affect all Arctic-related activities, including Arctic research. Thus, there is a need for a strong political will and an effective action plan.

On the issue of ways to collaborate between researchers to solve this problem on the international front, we should seriously consider the following issues:

(1) Improve the connections between the providers and users of Arctic (in terms of logistics);

(2) Organize more international workshops dedicated to the subject;

(3) Disclosure of international collaborators' research data, bearing in mind the difficulties regarding obtaining data from some countries and regions (e.g., Russia); and

(4) Have a comprehensive consideration of the benefits and support related to the involvement of non-Arctic countries in such collaboration and benefits for northern (indigenous) communities.

Conducting research and keeping it on the shelf or only making it available to the academic community further defeats the purpose of collaboration in providing comprehensive solution to the problems of climate change and the Arctic. Thus, there is an urgent need to convert research into an easily understandable format for use and understanding. Some of the modern ways include leveraging the use of social media, use videos, and newsletters, to name but a few. Finally, so far, unlike oil spills, we have not paid too much attention to the problem of BC release in the Arctic region. This needs to change as it plays a significant role in contributing to the environmental pollution in the region. Further research is required in this area.

5 Conclusion

Climate change remains a divisive topic in many parts of the world, and to many, it can only be negative. Flooding's impacts on the infrastructure and properties are some of these negatives. However, the Arctic countries have different perspectives on the climate change. To them, it is an opportunity to increase shipping activities, open communities up to tourism, and display the beauty of this unique part of the world. Despite these opportunities, those who have had the opportunity to work in the Arctic regions remain rather cautious about how we should and could explore these opportunities. Navigation equipment needs to be updated, while BC and oil spills need to be addressed holistically from global, national, and local perspectives.

The collected views on climate change effects and the Arctic have shown the need for a unique alliance to be forged so that there will be opportunities for social scientists, politicians, natural scientists, and other academics to sit down at the same table to discuss issues related to climate change and by extension the Arctic.

The inference from these views is that climate change-related issues are crucial and need urgent attention. In this case, it is pivotal to understand both the benefits and the negative consequences of climate change. Hitherto, many funding agencies are interested in funding climate change and Arctic-related research works. These agencies include both private, public, and international entities. However, we have to present it in a way that they can understand and appreciate the urgency and complexity of the tasks. However, many barriers exist due to different perspectives on the same problem. The need for social and natural scientists to work hand in hand has never been more critical than now.

References

Afenyo, M., Ng, A. K. Y., & Jiang, C. (2021). A multi-period model for assessing the socio-economic impact of oil spills in the Arctic. *Risk Analysis: An international Journal.*, *42*(3), 614–633.

Brigham, L. (2000). Polar icebreakers at the end of the twentieth century. *Polar Record*, *36* (198), 247–249.

Dawson, J., Maher, P. T., & Slocombe, S. D. (2007). Climate change, marine tourism, and sustainability in the Canadian Arctic: Contributions from systems and complexity approaches. *Tourism in Marine Environments*, *4*(2–3), 69–83.

Frolov, I. E. (2015). Development of the Russian Arctic zone: Challenges facing the renovation of transport and military infrastructure. *Studies on Russian Economic Development*, *26*, 561–566.

Gaston, A. J., & Bradstreet, M. S. W. (1993). Intercolony differences in the summer diet of thick-billed Murres in the eastern Canadian Arctic. *Canadian Journal of Zoology*, *71*, 9.

Høye, T., & Sikes, D. (2013). Arctic entomology in the 21st century. *The Canadian Entomologist*, *145*(2), 125–130.

Huddart, D., & Stott, T. (2020). Adventure tourism in the Canadian Arctic. In *Adventure tourism*. Cham: Palgrave Macmillan.

Jakubas, D., Wojczulanis-Jakubas, K., Iliszko, L. M., Strøm, H., & Stempniewicz, L. (2017). Habitat foraging niche of a high Arctic zooplanktivorous seabird in a changing environment. *Scientific Reports*, 7, 16203. https://doi.org/10.1038/s41598-017-16589-7.

Panahi, R., Afenyo, M., & Ng, A. K. Y. (2023). Developing a resilience index for safer and more resilient arctic shipping. *Maritime Policy & Management*, *50*(7), 861–875.

Schiøtt, S., Tejsner, P., & Rysgaard, S. (2022). Inuit and local knowledge on the marine ecosystem in Ilulissat Icefjord, Greenland. *Human Ecology*, *50*, 167–181.

Schumann, K., Völker, D., & Weinrebe, W. R. (2012). Acoustic mapping of the Ilulissat Ice Fjord mouth, West Greenland. *Quaternary Science Reviews*, *40*, 78–88.

Stephen, K., & Knecht, S. (2017). *Governing Arctic change - global perspectives*. The Arctic Institute. https://www.thearcticinstitute.org/governing-arctic-change/. (Accessed 9 September 2023).

Wilson Institute (WI). (2022). *Ukraine and the Arctic: Perspectives, Impacts, and Implications*. Hindsight Upfront Ukraine. https://www.wilsoncenter.org/hindsightukraine. (Accessed 9 September 2023).

Wråkberg, U. (2006). Nature conservationism and the Arctic commons of Spitsbergen 1900–1920. *Acta Borealia*, *23*(1), 1–23.

CHAPTER TWELVE

Cybersecurity in the Arctic: Are businesses and infrastructures safe?

1 Introduction

Several cyberattacks have taken place and resulted in the loss of money and critical assets and negative safety perception of the organizations affected. The Maersk attack in 2017 has cost the company approximately $300 million. The recent blockage of the Suez Canal, even though not related to a cybersecurity attack, costs the world economy 12% of global trade (Afenyo et al., 2022; Afenyo & Caesar, 2023). As of 2017, the major ports around the world were the subject of approximately 12 cyberattacks per day. On average, shipowners pay approximately $ 3.1 million in ransom due to maritime cyberattacks (Afenyo & Caesar, 2023).

Malware attacks and, in particular, ransomware attacks have risen to 13% in the last 5 years according to Nivedita (2023). The famous June cyberattack, which affected shipping terminals all over the world and affected the Port of Los Angeles costing between $200 and $300 million (Leovy, 2017), signals a rather alarming trend. While there are many vulnerable firms, the rate of successful attacks would be lower if measures are put in place.

Many ports in the US work closely with the USCG to not only secure borders but also build a resilient cybersecurity system. A 2023 USCG document shows that the United States and other maritime powers have a long way to go before they can properly deal with the problem (United States Coast Guard, 2023). One attack on a critical maritime infrastructure like causing the blockage of the Panama or Suez Canal can affect not just the economy but the timely delivery of essential commodities like medicine.

The Arctic, just like any maritime space, is not spared such occurrences. The risks associated with Arctic operations are based on rather complex and intertwined technologies available. Most operations are based on the

Safe and Sustainable Arctic Shipping Management and Development
https://doi.org/10.1016/B978-0-443-18849-7.00016-6
Copyright © 2024 Elsevier Inc.
All rights are reserved, including those for text and data mining, AI training, and similar technologies.

Internet and other digital infrastructure. Some of the mostly used digital systems include radar stations, GPS navigation, scanners, and others. These help in tracking rather complex systems in a simplistic way. Such technologies not only make operations efficient and easy but also help reduce the number of people required to run these systems. Digitalization comes with it consequences. One such is the vulnerability of the systems. Some of these systems can be hacked remotely and could be attacked with ransomware, malware, etc. Hacking activities used to be low-frequency, highly consequential events. These have changed lately moving to high-frequency and high-consequence events. In addition, these events are very unpredictable, and the inconsistent weather patterns these days even make it extremely difficult to put in place very reliable countermeasures to deal with such occurrences. Therefore, there is an urgent need to enact policies that not only are specific to dealing with maritime cyberattacks but also incorporate resilience principles (Trump, Hossain, & Linkov, 2020). This implies that the policies should encompass remedies for recovering the whole Arctic infrastructure systems related to the digital space to the original forms before the attack. This type of approach will enable stakeholders to have a comprehensive understanding of the problem related to Arctic cyberthreats and design appropriate strategies to deal with the issue.

Critical infrastructure, for example, the Kolos Data Center located in Norway, leverages the cold temperatures in the Arctic. However, with the huge amount of data to be stored there, it is very vulnerable to cyberattacks and should be subjected to in such a way that the consequences would be dire. Repairing after an attack would be extremely difficult to implement. If we imagine the computer system of a ship sailing in the Arctic hacked, where it will be extremely difficult to call for a backup within a short period of time, the likelihood that the people on board the ship may not survive is very high (Marchenko & Babyr, 2021).

While in many areas the Internet is built with backup nodes, this is not the same in the Arctic area. In the Arctic, however, many of the individual communities rely extensively on single fiber optic cables. Any problem with the cable implies a disruption of Internet services. Repair work also becomes extremely difficult because of the remote nature of Internet infrastructure in the Arctic. As an example, in 2018, a damaged submarine in the cable in the Arctic required 8 months to fix. The use of satellite communications, though a reasonable substitute, has its own challenges as well since these satellites are just stationed above the Arctic Circle. The last problem with Arctic connectivity is the fact that most of the Arctic territories are so-called semi-autonomous states. This means they are part of a major state, which has its own connectivity and cybersecurity policies, which may not necessarily be helpful to these regions with special characteristics than the others. These regions are also less populated, spread apart, and the environmental conditions are harsh, as well as the culture is also not homogeneous with each small settlement having its individual cultures. The services and authorities running these regions often find it difficult and costly to develop separate operations and services for just small numbers as opposed to larger groups (Marchenko & Babyr, 2021).

2 Arctic cybersecurity and business operations

Recently, a report was released by the University of Cambridge Centre for Risk Studies, which states that a single attack has major implications for ports, especially for ports in the Asia-Pacific. Damages could be up to approximately USD110 billion. The scenario for this to occur will be when about 15 ports are attacked across Japan, China, South Korea, Singapore, and Malaysia. Since Russia and China have made the Northern Sea Route a major route for transporting goods and LNG, Arctic ports face problems similar to those in open waters. In fact, the loss would be more should a similar incident occur in any of the active Arctic ports that are currently used for the delivery of items through the NSR.

Even more disturbing is the fact that the number of monthly attacks on major ports like the Port of Los Angeles is approximately 40 million, according to a statement issued by Mr. Gene Seroka, the Executive Director of the Port (Fenwick, 2022).

Business leaders are aware of the amount of cargo that is moved across the ocean and the importance of ocean shipments (Fitton et al., 2015). A significant disruption for any major player in the maritime industry could cause significant damage to that firm (Greenberg, 2018). Such a disruption would certainly have damaging ripple effects throughout the supply chain as a whole and for industries relying on timely shipments (Jensen, 2015). In recent years, the maritime transportation industry has undergone a digital transformation, which provides many benefits yet opens up opportunities for bad actors as new technology is implemented (Tijan et al., 2021). A crippling cyberattack on any maritime-related firm would have devastating consequences and leads to loss of money and reputation. There is simply no way to predict how long it will take for systems to be restored (Pownall, 2019). While there are many unknowns, there is one certainty. The complexity and frequency of these attacks are increasing and the Arctic conditions only make it worse.

One of the most well-known and cited cybersecurity attacks was the 2017 Maersk attack. This attack highlights some of the weaknesses still seen in the maritime industry today. Outdated systems, insufficient network segmentation, and other factors contributed to the successful attack. While the attack was not directed at them, the gaps in their cybersecurity practices left them vulnerable. Ultimately, Maersk was saved by good backup practice, allowing them to restore workstations and servers with some success rather than rebuilding them from scratch (Greenberg, 2018). This incident further hastened the realization of how important good cybersecurity practices were. Unfortunately, today, many firms remain vulnerable (Laursen, 2015). There is an ever-increasing frequency and complexity of attacks, and very few measures are being utilized to mitigate these new risks.

Like any other activities worldwide these days, Arctic shipping and oil and gas exploration and other related activities are prone to cyberattacks. These attacks often come in different forms, including cyber and piracy (Anonymous, 2019). Future increase in cyberattacks and cyber vulnerability for maritime supply chains is already an established consensus among practitioners and academics. In the Arctic, however, this is only evolving. Even for open waters, there is little information in the literature regarding the research gaps and future needs of the maritime industry in terms of cybersecurity.

Companies are already struggling to deal with maritime cybersecurity attacks for open waters (Afenyo et al., 2022). These attacks have serious consequences in terms of operations and economics. Arctic-specific cyberattacks are even more dire, mainly because of the harsh nature of the Arctic as well as its remoteness. Cybersecurity in this region is key to protecting critical assets and operations as well as facilitating commerce. Cybersecurity issues in the Arctic in the context of business operations can be examined in terms of operations management, training of personnel, policies and regulations, potential economic losses, data management, and education. To address the remoteness of Arctic infrastructure especially oil and gas platforms in Russia, Marchenko and Babyr (2021) proposed the design and use of an Intelligent Digital Arctic Logistics Management System. The system would involve the use of artificial intelligence to develop, produce, and transport oil from remote Arctic regions to the points of interest without any human interference. One of the advantages of such a system would be its efficiency, resilience, and long-term cost cutting. However, such a system would largely depend on the availability of reliable data, which is often a very big problem when it comes to Arctic operations and infrastructure. Already, several organizations are not very comfortable going ahead with Arctic shipping as a cheap means of transporting goods (Nilsen, 2020). Companies, such as Ralph Lauren Corporation, Kuehne + Nagel, PUMA, H&M Columbia, and Gap Inc. have all subscribed to this thinking. However, in places like Russia, the opposite is happening. Oil and gas operations, shipping, and other Arctic-related activities have increased over the years, most of which use a lot of cyberinfrastructure (Franiok, 2020). In this case, it is extremely important that cybersecurity should be considered from the conceptual stage of engineering design with a plan for seamless collaboration among major players.

Currently, it is easy to predict where the ship's position is going to be and so extremely vulnerable to attacks. Most organizations are behind with respect to realigning their structures to better reflect the realities of cyber threats in the Arctic region. While business operations have picked up in some parts of the Arctic, one major issue remains, that is the right training for the personnel who operate Arctic infrastructure. Currently, most of the maritime academies are focused on Arctic maritime cybersecurity training. What this means is that most graduates are not equipped with the right knowledge and skills to deal with Arctic maritime cyber threats when they are faced with one. It would therefore be very important to design effective training programs to adequately equip maritime professionals. This is only possible with a thorough understanding of the psyche of the participants. Training workers in private and government institutions to readily deal with maritime cybersecurity threats using a virtual environment is one of the best ways to deal with this issue. A strong case for paradigm shifts in the way maritime cybersecurity education is conducted is also necessary. The curriculum needs to be updated to better reflect the new changes in the field.

Every business operation requires the right regulations for efficient and effective operations. However, in the case of Arctic maritime cyber regulations, this does not exist. The most relevant regulations are the Maritime Transportation Security Act (MTSA) and International Maritime Organizations' (IMO) International Ship and Port Facility Security (ISPS) Code, which do not address cybersecurity. In a report released by the United States Department of Navy, they identified Arctic maritime cybersecurity as very important in ensuring that military as well as civilian activities are safe. For example, the report pointed out that increasingly there is access to new Arctic undersea fiber optic cable routes linking Europe, Asia, and North America. With this information, these fiberoptic cables have become major targets for numerous cyberattacks. Other sectors susceptible to such attacks include but are not limited to shipbuilding, energy, and shipping sectors. At the top of the list of the most hacked enterprises are the research and development communities (Anonymous, 2021; Tallis, 2021). Finally, there is limited documentation with regard to how much each of the notable attacks cost internally. While the media reports estimates, companies are coy on the real cause and cost associated with such attacks.

3 Data and Arctic maritime cybersecurity

It is difficult to get any planning done without the right quantity and quality of data. A lot of companies are very protective of corporate data. Even though some public databases exist, for example, the Norwegian Shipowners' Mutual War Risk Insurance Association database and the Norwegian Maritime Cyber Resilience Centre, they are limited in terms of which data they can collect. They only have permission to collect much data from certain countries. Most countries are behind in this regard, even though some data have been collected over the years. Remoteness and the harsh environmental conditions further north remain a major challenge to data collection and therefore there is still a lot of missing information on the Arctic region making it difficult to design an efficient cyber-physical infrastructure for these areas. Most of the comprehensive data collected so far have been by the military, both by that of the United States and Russia. However, the bulk of the data is classified and hence not accessible to the public (Heyn et al., 2018).

If we are to design effective cybersecurity protective systems, we will need adequate data to make this happen. Data here should be focused not only on the Internet but also on the geology of the region. This could give us an idea of where we could install or locate some of the more sensitive cyber-physical infrastructures in the Arctic. There should also be collaboration among the Arctic countries to share data so as to help facilitate the building of an Arctic database system.

4 Risk, insurance, and Arctic maritime cybersecurity

The more disturbing part of the study alluded to earlier on Asian ports, according to the *Cyber Risk Management Project (CyRiM)* conducted by one of Lloyd's partners, states that approximately 92% of these economic losses due to cyberattacks are not insured (Yu, 2019). This is particularly disturbing considering how intertwined the activities of ports around the globe are related. The big question therefore is, what is the main reason for this? Could it be that it is just difficult for insurance companies to insure such events or companies are just not interested and probably do not see any needs for it?

Attacks on systems occur through many different vectors. Most, but not all these attacks can be mitigated through implementing various measures to lower the risk. One way of protecting a business financially is through insurance. This is important for the maritime transportation industry because of the amount of capital it deals in and how large assets can quickly become compromised. Today, there are a few firms that offer full coverage for cybersecurity attacks. This is quite possible because the knowledge surrounding the nature of cyberattacks and the potential financial consequences is lacking. Additionally, this is also because there are so many factors affecting the size and scale of the attack if successful and the damage caused.

The literature published previously on the subject yielded a myriad of findings on the themes of maritime cybersecurity insurance generally and Arctic maritime cybersecurity, in particular. Some of the major findings show the need to examine the importance of protecting such a large critical industry and discuss the potential consequences. Many others discussed the legal aspects and wording of contracts regarding specific insurance policies over maritime cyberattacks. Table 1 presents a snapshot of some of the other studies related to the subject matter.

Table 1 Studies reviewed that are related to maritime cybersecurity.

Author	Focus	Conclusion
Kao (2021)	Using insurance to mitigate cybersecurity (CS) risks. Implied warranty of seaworthiness and proximate clause in maritime law can help determine what is covered.	Specific language can be used in insurance policies to be proactive in risk management.
Leffingwell and Palmer (2022)	Introduces practices such as easily understood incident response policies, improved training for staff, and collaboration to help prevent and mitigate CS threats at early stages.	Gives readers simple practical advice to help mitigate the risks of human error and improve CS policies and practices.

Continued

Table 1 Studies reviewed that are related to maritime cybersecurity—cont'd

Author	Focus	Conclusion
Sullivan and Karamperidis (2021)	States cybercriminals know how vulnerable and outdated the systems are in the maritime industry. In addition, it addresses outdated terminology in insurance policies.	Threats are constantly evolving and arise in many different forms. New bad actors and new legal challenges arise in maritime cybersecurity insurance.
Weaver et al. (2022)	Used simulated cyberattacks to evaluate the flaws in IT/OT systems and then recorded estimated damages. The increase in automation allows new risks to be considered.	Risk assessments must evaluate the interactions between IT and OT systems.

However, when it comes to Arctic maritime cyber insurance, there is next to zero information. In fact, most maritime cyberattacks are not insured making structures in the maritime domain a high risk. To put the risk of maritime cyber incidents into perspective, just the first quarter of 2022 alone, cyberattacks on maritime systems and related infrastructure increased by 28% (Violino, 2022). Cyber-insecurity not only poses increased risks to maritime operations, but also to general planning, which more and more has begun to rely on cyber insurance for predictable outcomes. According to BlackBerry, about 60% of businesses surveyed said they would reconsider making an agreement with a vendor if they learned that a vendor lacked comprehensive cyber-risk coverage. A report by AM best indicates that cybersecurity premiums collected reached a high of USD7.2 billion (Price, 2023). This only shows one thing, that cyber incidents are on the ascendency. A lot of insurance companies do not have the required knowledge to properly evaluate the potential consequences of Arctic maritime cyberattacks should they happen. Recently, the Swedish Club has engaged experts in the field to help develop a framework to cover cyberattacks involving its members. Some of the coverages include emergency response, physical damage, loss of hire, and wreck removal (Wingrove, 2023), which are scheduled to commence in 2024. This is one of the first of its kind in the world to be done according to the guidelines of the International Maritime Organizations' (IMO) Cybersecurity. The guidelines recognize that no maritime infrastructure is the same and the environments are also very different. Therefore, they are designed to cater to different environments and infrastructures. They are applied to all the organizations in the shipping industry for safe and efficient shipping (International Maritime Organization (IMO), 2022).

5 Discussions

The Arctic region is home to many data hubs, as alluded to earlier as well as huge energy resources that largely rely on secure computer control. While complete elimination of all cyber-related threats could not be achieved, a good resilient system might help. The people, including indigenous communities, in the Arctic are not left out of digital evolution. Indeed, there is a worry about the protection of traditional culture as indigenous communities get exposed to the mainstream media. With digitization comes vulnerability to the Arctic cyberinfrastructure, remote and digitally controlled technologies in a harsh, rapidly changing region. It is not going to be easy. The security of ports, ships, offshore oil, and gas platforms also needs to be looked at critically in terms of how they can be protected in the face of these cyberattacks. In the Arctic region, the population is very sparse and information infrastructure is normally much weaker than in the nonpolar regions. Internet connectivity in these regions is extremely difficult and is often done in a different way and very expensive (Anonymous, 2019). Also, the ongoing Russo-Ukrainian conflict has created a major problem for the Arctic nations since Russia has been sanctioned by many Arctic countries. It means that collaborations for cybersecurity are unlikely to happen in the foreseeable future, hence exposing the region's infrastructures to further attacks from bad actors, individuals, and other governments.

In addition, as argued by Hossain and Klein (2020), cybersecurity issues in the Arctic region should be looked at through the lens of human beings, especially as we have the effect of climate change all around the Arctic. We cannot detach policies dealing with cybersecurity from climate change, and for that matter, the humans that inhabit these regions.

6 Conclusion

Arctic maritime cybersecurity is a relatively new topic in Arctic studies. However, the problems associated with maritime cyberspace are not peculiar to the industry. The evaluation conducted in the chapter shows that the cyberinfrastructure in the Arctic is very vulnerable to attack, just like in any other industry. Nonetheless, there is the challenge of offering insurance for these developments. This, however, presents another challenge considering the fact that data are scarce on the occurrence of the attacks even in the nonpolar regions, notwithstanding of the fact that some insurance companies have started venturing into this domain. The only very certain thing is that as the number of attacks increases, the premiums will also increase. It is important to have a resilient cybersecurity system to ensure that businesses can operate in the region without any fear. Moving forward, there is a need for the Arctic countries to build a database and probably an Arctic cyberforce for individual countries for the time being and then later consider expanding it to include a collaboration of the Arctic nations.

References

Afenyo, M., & Caesar, D. (2023). Maritime cybersecurity threats: Gaps and directions for future research. *Ocean and Coastal Management, 236*, 106493.

Afenyo, M., Ng, A. K. Y., & Jiang, C. (2022). A Bayesian-loss function (BLF) hybrid model for assessing the marine liability regime for ship-source oil spills in the Arctic. *Risk Analysis, 43*(8), 1557–1571. https://doi.org/10.1111/risa.14025.

Anonymous. (2019). *Resilience is the key to cybersecurity in the Arctic.* Arctic Centre, University of Lapland. https://www.arcticfinland.fi/news/Resilience-is-the-key-to-cybersecurity-in-the-Arctic-/x1xawhz5/a8bf6b22-a504-468e-a8f7-ab4727c9cc9c#:~:text=Top%20scholars%20are%20seeking%20solutions,the%20Arctic%20against%20cyber%20threats. (Accessed 21 November 2023).

Anonymous. (2021). *Department Of the Navy: A Strategic Blueprint for the Arctic.* https://media.defense.gov/2021/Jan/05/2002560338/-1/-1/0/ARCTIC%20BLUEPRINT%202021%20FINAL.PDF/ARCTIC%20BLUEPRINT%202021%20FINAL.PDF. (Accessed 25 November 2023).

Fenwick, S. (2022). *Cyber-attacks on Port of Los Angeles have doubled since pandemic.* BBC News.

Fitton, O., Prince, D., Germond, B., & Lacy, M. (2015). *The future of maritime cyber Security.* Retrieved from Lancaster University.

Franiok, N. (2020). *Russian Arctic military bases.* American Security Project. https://www.americansecurityproject.org/russian-arctic-military-bases/ (Accessed 25 November 2023).

Greenberg, A. (2018). *The untold story of NotPetya, the most devastating cyberattack in history.* August 22. Retrieved from Wired. https://www.wired.com/story/notpetya-cyberattack-ukraine-russia-code-crashed-the-world/.

Heyn, B., Gosselin, J., & Uppalapati, S. (2018). *Data acquisition in the Arctic – A challenge throughout time.* NOR-R-AM. www.norramarctic-wordpress.com/2018/09/30/data-acquisition-in-the-arctic-achallenge throughout time/. (Accessed 27 November 2023).

Hossain, K., & Klein, J. (2020). Conceptualizing human-centric cyber security in the Arctic in light of digitalization and climate change. *Arctic Review on Law and Politics.* https://doi.org/10.23865/arctic.v11.1936.

International Maritime Organization (IMO). (2022). *Guidelines on maritime cyber risk management.* https://wwwcdn.imo.org/localresources/en/OurWork/Security/Documents/MSC-FAL.1-Circ.3-Rev.2%20-%20Guidelines%20On%20Maritime%20Cyber%20Risk%20Management%20(Secretariat)%20(1).pdf. (Accessed 27 November 2023).

Jensen, L. (2015). Challenges in maritime cyber-resilience. *Technology Innovation Management Review, 5*(4), 35–38.

Kao, M. (2021). Cybersecurity in the shipping industry and English marine insurance law. *Tulane Maritime Law Journal, 45*(3), 467–508.

Laursen, W. (2015). *Many operators open to simple cyber attack.* The Maritime Executive. June 10. Retrieved from https://maritime-executive.com/article/many-operators-open-to-simple-cyber-attack.

Leffingwell, M., & Palmer, P. A. (2022). Navigating cybersecurity preparedness – incident response in the maritime industry. *Tort Trial & Insurance Practice Law Journal, 57*(2), 1–15.

Leovy, J. (2017). Cyberattack cost Maersk as much as $300 million and disrupted operations for 2 weeks. *Los Angeles Times.* https://www.latimes.com/business/la-fi-maersk-cyberattack-20170817-story.html. (Accessed 1 September 2023).

Marchenko, R., & Babyr, A. (2021). Digitalization of Arctic logistics management systems for oil transportation. *Transportation Research Procedia, 54,* 953–960. https://doi.org/10.1016/j.trpro.2021.02.150.

Nilsen, T. (2020). Arctic shipping boycott gains support among big businesses, but not all are happy. *The Barents Observer.* https://thebarentsobserver.com/en/arctic/2020/01/avoiding-arctic-sea-routes-might-be-good-intention-isnt-solving-problem. (Accessed 25 November 2023).

Nivedita, J. (2023). *100+ ransomware attack statistics 2023: Trends and cost. Security Audit.* Astra. https://www.getastra.com/blog/security-audit/ransomware-attack-statistics/. (Accessed 1 September 2023).

Pownall, C. (2019). *The context and impact of Maersk's NotPetya cyber attack.* Research Gate.

Price, G. F. (2023). *Insights into the changing landscape of cyber insurance.* Frost Brown Todd Attorneys. https://frostbrowntodd.com/insights-into-the-changing-landscape-of-cyber-insurance/. (Accessed 27 November 2023).

Sullivan, B., & Karamperidis, S. (2021). Ports, Maritime & Waterways Forum: Cyber-risk in the ports and maritime sector. *Focus (2756–1755), 23*(1), 67.

Tallis, J. (2021). *A new Arctic strategy for an emerging maritime domain.* Center for International Maritime Security (SIMSEC). https://cimsec.org/a-new-arctic-strategy-for-an-emerging-maritime-domain/. (Accessed 25 November 2023).

Tijan, E., Jović, M., Aksentijević, S., & Pucihar, A. (2021). Digital transformation in the maritime transport sector. *Technological Forecasting and Social Change, 170.* Retrieved from https://www.sciencedirect.com/science/article/pii/S0040162521003115.

Trump, B. D., Hossain, K., & Linkov, I. (2020). *Cybersecurity and resilience in the Arctic.* IOS Press.

United States Coast Guard. (2023, January). *Maritime cybersecurity assessment and annex guide (MCAAG).* Retrieved from uscg.mil: https://www.dco.uscg.mil/Our-Organization/Assistant-Commandant-for-Prevention-Policy-CG-5P/Inspections-Compliance-CG-5PC-/Office-of-Port-Facility-Compliance/Domestic-Ports-Division/cybersecurity/.

Violino, B. (2022). *Rising premiums, more restricted cyber insurance coverage poses big risk for companies.* CNBC. https://www.cnbc.com/2022/10/11/companies-are-finding-it-harder-to-get-cyber-insurance-.html. (Accessed 28 November 2023).

Weaver, G., Feddersen, B., Marla, L., Wei, D., Rose, A., & van Moer, M. (2022). Estimating economic losses from cyber-attacks on shipping ports: An optimization-based approach. *SSRN Electronic Journal.* https://doi.org/10.2139/ssrn.3816659.

Wingrove, M. (2023). *Maritime cyber insurance offered to shipowners.* Riviera https://www.rivieramm.com/news-content-hub/news-content-hub/marine-insurance-offered-to-cover-cyber-attacks-78653. (Accessed 27 November 2023).

Yu, E. (2019). *One cyber-attack can cost major APAC ports $110B.* ZDNET.

Index

Note: Page numbers followed by *f* indicate figures and *t* indicate tables.

B

C

L

M

N

O

S

T

U

V

W

Y

9780443188497